In Situ Bioremediation

When does it work?

Committee on In Situ Bioremediation

Water Science and Technology Board

Commission on Engineering and Technical Systems

National Research Council

NATIONAL ACADEMY PRESS
Washington, D.C. 1993

National Academy Press • 2101 Constitution Avenue, N.W. • Washington, D.C. 20418

NOTICE: The project that is the subject of this report was approved by the Governing Board of the National Research Council, whose members are drawn from the councils of the National Academy of Sciences, the National Academy of Engineering, and the Institute of Medicine. The members of the committee responsible for the report were chosen for their special competences and with regard for appropriate balance.

This report has been reviewed by a group other than the authors according to procedures approved by a Report Review Committee consisting of members of the National Academy of Sciences, the National Academy of Engineering, and the Institute of Medicine.

Support for this project was provided by the U.S. Environmental Protection Agency under Agreement No. CR 820730-01-0, the National Science Foundation under Agreement No. BCS-9213271, the Electric Power Research Institute under Agreement No. RP2879-26, the Gas Research Institute, the American Petroleum Institute, Chevron USA, Inc., and the Mobil Oil Corporation.

Library of Congress Cataloging-in-Publication Data

In situ bioremediation / Water Science and Technology Board,
Commission on Engineering and Technical Systems, National Research Council.
p. cm.
Includes bibliographical references and index.
ISBN 0-309-04896-6
1. In situ bioremediation—Evaluation. I. National Research Council (U.S.). Water Science and Technology Board.
TD192.5.I53 1993 93-5531
628.5′2—dc20 CIP

B-184

Cover art by Y. David Chung. Title design by Rumen Buzatov. Chung and Buzatov are graduates of the Corcoran School of Art in Washington, D.C. Chung has exhibited widely throughout the country, including at the Whitney Museum in New York, the Washington Project for the Arts in Washington, D.C., and the Williams College Museum of Art in Williamstown, Massachusetts.

In brilliant colors, the cover art shows the amazing variety of unusual shapes found in bacterial life forms.

Printed in the United States of America

COMMITTEE ON IN SITU BIOREMEDIATION

BRUCE E. RITTMANN, *Chair*, Northwestern University, Evanston, Illinois
LISA ALVAREZ-COHEN, University of California, Berkeley
PHILIP B. BEDIENT, Rice University, Houston, Texas
RICHARD A. BROWN, Groundwater Technology, Inc., Trenton, New Jersey
FRANCIS H. CHAPELLE, U.S. Geological Survey, Columbia, South Carolina
PETER K. KITANIDIS, Stanford University, Stanford, California
EUGENE L. MADSEN, Cornell University, Ithaca, New York
WILLIAM R. MAHAFFEY, ECOVA Corporation, Redmond, Washington
ROBERT D. NORRIS, Eckenfelder, Inc., Nashville, Tennessee
JOSEPH P. SALANITRO, Shell Development Company, Houston, Texas
JOHN M. SHAUVER, Michigan Department of Natural Resources, Lansing, Michigan
JAMES M. TIEDJE, Michigan State University, East Lansing, Michigan
JOHN T. WILSON, Robert S. Kerr Environmental Research Laboratory, Ada, Oklahoma
RALPH S. WOLFE, University of Illinois, Urbana

Staff

JACQUELINE A. MACDONALD, Study Director
GREGORY K. NYCE, Senior Project Assistant
GREICY AMJADIVALA, Project Assistant
WYETHA TURNEY, Word Processor
KENNETH M. REESE, Editorial Consultant
BARBARA A. BODLING, Editorial Consultant

WATER SCIENCE AND TECHNOLOGY BOARD

Staff

COMMISSION ON ENGINEERING AND TECHNICAL SYSTEMS

Staff

The National Academy of Sciences is a private, nonprofit, self-perpetuating society of distinguished scholars engaged in scientific and engineering research, dedicated to the furtherance of science and technology and to their use for the general welfare. Upon the authority of the charter granted to it by the Congress in 1863, the Academy has a mandate that requires it to advise the federal government on scientific and technical matters. Dr. Bruce M. Alberts is president of the National Academy of Sciences.

The National Academy of Engineering was established in 1964, under the charter of the National Academy of Sciences, as a parallel organization of outstanding engineers. It is autonomous in its administration and in the selection of its members, sharing with the National Academy of Sciences the responsibility for advising the federal government. The National Academy of Engineering also sponsors engineering programs aimed at meeting national needs, encourages education and research, and recognizes the superior achievements of engineers. Dr. Robert M. White is president of the National Academy of Engineering.

The Institute of Medicine was established in 1970 by the National Academy of Sciences to secure the services of eminent members of appropriate professions in the examination of policy matters pertaining to the health of the public. The Institute acts under the responsibility given to the National Academy of Sciences by its congressional charter to be an adviser to the federal government and, upon its own initiative, to identify issues of medical care, research, and education. Dr. Kenneth I. Shine is president of the Institute of Medicine.

The National Research Council was organized by the National Academy of Sciences in 1916 to associate the broad community of science and technology with the Academy's purposes of furthering knowledge and advising the federal government. Functioning in accordance with general policies determined by the Academy, the Council has become the principal operating agency of both the National Academy of Sciences and the National Academy of Engineering in providing services to the government, the public, and the scientific and engineering communities. The Council is administered jointly by both Academies and the Institute of Medicine. Dr. Bruce M. Alberts and Dr. Robert M. White are chairman and vice chairman, respectively, of the National Research Council.

Preface

Bioremediation is a technology that is gaining momentum in technical, policy, and popular circles. It also is a technology associated with mystery, controversy, and "snake oil salesmen." When a representative of the U.S. Environmental Protection Agency suggested in the fall of 1991 that the Water Science and Technology Board conduct a study on bioremediation, it converged with the board's internal initiative to "do something" in the area. Several high-quality workshops and conferences had occurred in the previous year that generated publications describing what is needed for bioremediation to fulfill its potential. The board needed to design a study that would do more than repeat what was already available, that would be completed in a time frame commensurate with the urgent needs of those involved in bioremediation, and that would meet the high standards expected of the National Academy of Sciences. These criteria inevitably led to the subject of this report and to a unique format for conducting the study.

The study's subject—"In Situ Bioremediation: When Does It Work?"—narrows the focus to two critical facets of bioremediation. First, it addresses the use of microorganisms to remove contamination from ground water and soils that remain in place (i.e., in situ) during the cleanup. This focus distinguishes in situ bioremediation of the subsurface from significantly different applications of bioremediation, such as to treat oil tanker spills, wastewaters, or sludges. Second, the primary object of the study is to provide guidance on how to evaluate when an in situ bioremediation process is working or has worked. This focus is most important because the in situ environment is highly complex and very difficult to observe. Therefore, tools from several scientific and engineering disciplines must be used in a sophisticated manner if the success of a bioremediation effort is to be evaluated. Guidance is acutely needed today because most people faced with making decisions about bioremediation projects do not have the interdisciplinary knowledge to integrate all of the necessary tools.

The format for this study was unique and designed to meet two criteria: meaningful interdisciplinary interchange and timeliness. To gain interchange, a committee of 14 was carefully chosen to include recognized leaders in academic research, field practice, regulation, and industry. A balance was achieved between those involved in research fundamentals and those involved in the practical aspects of application, as well as between scientists and engineers. Once the committee of interdisciplinary experts was assembled, meaningful interchange was fostered by an intensive week-long workshop at the National Research Council. The goals were to maximize opportunities for formal and informal interchange among the committee members and to build a common purpose. Both goals were achieved, directly leading to a consensus about the issues and what were to be the committee's recommendations.

Timeliness was a prime consideration in designing the study's format. In order to accelerate interdisciplinary communications, nine committee members prepared seven background papers in advance of the week-long workshop. At the workshop, the committee initially generated its own discussion topics and then systematically discussed them. Key to timeliness and keeping the committee "on target" was preparation of a draft report during the workshop. Near the end of the workshop, the committee reviewed the draft report, which refocused the entire group on exactly what it wanted to say.

Appearing first in this volume is the committee's report, which describes the principles and practices of in situ bioremediation and provides practical guidelines for evaluating success. The report's guidelines should be immediately useful to regulators, practitioners, and buyers who are involved in decision-making processes involving bioremediation. We envision that the report will provide a commonly accepted basis for which all parties can agree to specific evaluation protocols. Also included here are the seven background papers. These papers will give the reader added insight into the different perspectives that were brought to the committee. The entire report has been reviewed by a group other than the authors, but only the committee report was subjected to the report review criteria established by the National Research Council's Report Review Committee. The background papers have been reviewed for factual correctness.

Special acknowledgment must go to several individuals who contributed to the committee's overall effort in special ways. First, Dick Brown and Jim Tiedje joined me on the executive committee, which had the all-important tasks of identifying and recruiting committee members and which also oversaw the committee's management. Second, Eugene Madsen, the committee's rapporteur, wrote the first draft of the report during the workshop and prepared an excellent second draft after the workshop. Eugene did these crucial and grueling tasks with skill and good humor. Finally, Jackie MacDonald, staff officer for the committee, made this unique effort possible. She efficiently arranged all the logistics for the workshop and for publishing the book. Even more importantly, she used her exceptional technical and editorial skills to ensure that the report and the background papers are logical, correct, understandable, and interesting to read. The committee members owe Jackie a debt of gratitude for making us sound more intelligent and better organized than we might actually be.

Finally, I want to mention two possible spin-off benefits of the study and report. First, most of the principles and guidelines described here also apply to evaluating bioremediation that does not occur in situ. Although the inherent difficulties of working in an in situ environment make evaluation especially challenging, other bioremediation applications also are subject to uncertainties and controversy that can be resolved only with the kind of rational evaluation strategies described here. Second, the format for the workshop might provide a prototype for effective interdisciplinary communications, one of the most critical needs for implementing bioremediation, as well as other technologies.

Bruce E. Rittmann, *Chair*
Committee on In Situ Bioremediation

Contents

In Situ Bioremediation

When does it work?

Executive Summary

The United States is investing billions of dollars in cleaning up polluted ground water and soils, yet this large investment may not be producing the benefits that citizens expect. Recent studies have revealed that because of limitations of ground water cleanup technologies, there are almost no sites where polluted ground water has been restored to a condition fit for drinking. While soil cleanup efforts have come closer to meeting regulatory goals, the technologies typically used to decontaminate soils often increase the exposure to contaminants for cleanup crews and nearby residents.

The limitations of conventional ground water cleanup technologies and the hazards of conventional soil treatment methods—along with the high costs of both—have spurred investigations into alternative cleanup technologies, including in situ bioremediation. In situ bioremediation uses microorganisms to destroy or immobilize contaminants in place. The technology already has achieved a measure of success in field tests and commercial-scale cleanups for some types of contaminants.

Proponents of in situ bioremediation say the technology may be less costly, faster, and safer than conventional cleanup methods. Yet despite mounting evidence in support of the technology, bioremediation is neither universally understood nor trusted by those who must approve its use. Bioremediation is clouded by controversy over what it does and how well it works, partly because it relies on microorganisms, which cannot be seen, and partly because it has become attractive for "snake oil salesmen" who claim to be able to solve all types of contamination problems. As long as the controversy remains, the full potential of this technology cannot be realized.

In this report the Committee on In Situ Bioremediation communicates the scientific and technological bases for in situ bioremediation, with the goal of eliminating the mystery that shrouds this highly multidisciplinary technology. The report presents guidelines for evaluating in situ bioremediation projects to determine whether they will or are meeting cleanup goals. The Committee on In Situ Bioremediation was established in June 1992 with the specific task of developing such guidelines, and it represents the span of groups involved in bioremediation: buyers of bioremediation services, bioremediation contractors, environmental regulators, and academic researchers. Included with the report are seven background papers, authored by committee members, representing the range of perspectives from which bioremediation may be viewed.

PRINCIPLES OF BIOREMEDIATION

The most important principle of bioremediation is that microorganisms (mainly bacteria) can be used to destroy hazardous contaminants or transform them to less harmful forms. The microorganisms act against the contaminants only when they have access to a variety of materials—compounds to help them generate energy and nutrients to build more cells. In a few cases the natural conditions at the contaminated site provide all the essential materials in large enough quantities that bioremediation can occur without human intervention—a process called *intrinsic bioremediation*. More often, bioremediation requires the construction of engineered systems to supply microbe-stimulating materials—a process called *engineered bioremediation*. Engineered bioremediation relies on accelerating the desired biodegradation reactions by encouraging the growth of more organisms, as well as by optimizing the environment in which the organisms must carry out the detoxification reactions.

A critical factor in deciding whether bioremediation is the appropriate cleanup remedy for a site is whether the contaminants are susceptible to biodegradation by the organisms at the site (or by organisms that could be successfully added to the site). Although existing microorganisms can detoxify a vast array of contaminants, some compounds are more easily degraded than others. In general, the compounds most easily degraded in the subsurface are petroleum hydrocarbons, but technologies for stimulating the growth of organisms to degrade a wide range of other contaminants are emerging and have been successfully field tested.

The suitability of a site for bioremediation depends not only on the contaminant's biodegradability but also on the site's geological and chemical characteristics. The types of site conditions that favor bioremediation differ for intrinsic and engineered bioremediation. For intrinsic bioremediation, the key site characteristics are consistent ground water flow throughout the seasons; the presence of minerals that can prevent pH changes; and high concentrations of either oxygen, nitrate, sulfate, or ferric iron. For engineered bioremediation, the key site characteristics are permeability of the subsurface to fluids, uniformity of the subsurface, and relatively low (less than 10,000 mg/kg solids) residual concentrations of nonaqueous-phase contaminants.

When deciding whether a site is suitable for bioremediation, it is important to realize that no single set of site characteristics will favor bioremediation of all contaminants. For example, certain compounds can only be degraded when oxygen is absent, but destruction of others requires that oxygen be present. In addition, one must consider how the bioremediation system may perform under variable and not perfectly known conditions. A scheme that works optimally under specific conditions but poorly otherwise may be inappropriate for in situ bioremediation.

THE CURRENT PRACTICE OF BIOREMEDIATION

Few people realize that in situ bioremediation is not really a "new" technology. The first in situ bioremediation system was installed 20 years ago to clean up an oil pipeline spill in Pennsylvania, and since then bioremediation has become well developed as a means of cleaning up easily degraded petroleum products. What is new is the use of in situ bioremediation to treat compounds other than easily degraded petroleum products on a commercial scale. The principles of practice outlined here were developed to treat petroleum-based fuels, but they will likely apply to a much broader range of uses for bioremediation in the future.

Engineered Bioremediation

Engineered bioremediation may be chosen over intrinsic bioremediation because of time and liability. Where an impending property transfer or potential impact of contamination on the local community dictates the need for rapid pollutant removal, engineered bioremediation may be a more appropriate remedy than intrinsic bioremediation. Because engineered bioremediation accelerates biodegradation reaction rates, it requires less time than intrinsic bioremediation. The shorter time requirements reduce the liability for costs required to maintain and monitor the site.

Since many petroleum hydrocarbons require oxygen for their degradation, the technological emphasis of engineered bioremediation systems in use today has been placed on oxygen supply. Bioremediation systems for soil above the water table usually consist of a set of vacuum pumps to supply air (containing oxygen) and infiltration galleries, trenches, or dry wells to supply moisture (and sometimes specific nutrients). Bioremediation systems for ground water and soil below the water table usually consist of either a set of injection and recovery wells used to circulate oxygen and nutrients dissolved in water or a set of compressors for injecting air. Emerging applications of engineered bioremediation, such as for degradation of chlorinated solvents, will not necessarily be controlled by oxygen. Hence, the supply of other stimulatory materials may require new technological approaches even though the ultimate goal, high biodegradation rates, remains the same.

Intrinsic Bioremediation

Intrinsic bioremediation is an option when the naturally occurring rate of contaminant biodegradation is faster than the rate of contaminant migration. These relative rates depend on the type and concentration of contaminant, the microbial community, and the subsurface hydrogeochemical conditions. The ability of native microbes to metabolize the contaminant must be demonstrated either in field tests or in laboratory tests performed on site-specific samples. In addition, the effectiveness of intrinsic bioremediation must be continually monitored by analyzing the fate of the contaminants and other reactants and products indicative of biodegradation.

In intrinsic bioremediation the rate-controlling step is frequently the influx of oxygen. When natural oxygen supplies become depleted, the microbes may not be able to act quickly enough to contain the contamination. Lack of a sufficiently large microbial population can also limit the cleanup rate. The microbial population may be small because of a lack of nutrients, limited availability of contaminants resulting from sorption to solid materials or other physical phenomena, or an inhibitory condition such as low pH or the presence of a toxic material.

Integration of Bioremediation with Other Technologies

Bioremediation frequently is combined with nonbiological treatment technologies, both sequentially and simultaneously. For example, when soil is heavily contaminated, bioremediation may be implemented after excavating soils near the contaminant source—a process that reduces demand on the bioremediation system and the immediate potential for ground water contamination. Similarly, when pools of contaminants are floating on the water table, these pools may be pumped to the surface before bioremediation of residual materials. Bioremediation may follow treatment of the ground water with a conventional pump-and-treat system designed to shrink the contaminant plume to a more manageable size. Bioremediation may also be combined with a vapor recovery system to extract volatile contaminants from soils. Finally, it is possible to follow engineered bioremediation, which cleans up most of the contamination, with intrinsic bioremediation, which may be used for final polishing and contaminant containment.

EVALUATING IN SITU BIOREMEDIATION

The inherent complexity of performing bioremediation in situ means that special attention must be given to evaluating the success of a project. The most elemental criterion for success of an in situ bioremediation effort is that the microorganisms are mainly responsible for the cleanup. Without evidence of microbial involvement, there is no way to verify that the bioremediation project was actually a bioremediation—that is, that the contaminant did not simply volatilize, migrate off site, sorb to the soil, or change form via abiotic chemical reactions. Simply showing that microbes grown in the lab have the *potential* to degrade the contaminant is not enough. While bioremediation often is possible in principle, the more relevant question is, "Are the biodegradation reactions actually occurring under site conditions?"

No one piece of evidence can unambiguously prove that microorganisms have cleaned up a site. Therefore, the Committee on In Situ Bioremediation recommends an evaluation strategy that builds a consistent, logical case for bioremediation based on converging lines of independent evidence. The strategy should include three types of information:

1. documented loss of contaminants from the site,
2. laboratory assays showing that microorganisms from site samples

have the *potential* to transform the contaminants under the expected site conditions, and

3. one or more pieces of information showing that the biodegradation potential is *actually realized* in the field.

Every well-designed bioremediation project, whether a field test or full-scale system, should show evidence of meeting the strategy's three requirements. Regulators and buyers of bioremediation services can use the strategy to evaluate whether a proposed or ongoing bioremediation project is sound; researchers can apply the strategy to evaluate the results of field tests.

The first type of evidence—documented loss of contaminants from the site—is gathered as part of the routine monitoring that occurs (or should occur) at every cleanup site. The second type of evidence requires taking microbes from the field and showing that they can degrade the contaminant when grown in a well-controlled laboratory vessel. The most difficult type of evidence to gather is the third type—showing that microbes in the field are actively degrading the contaminant. There are two types of sample-based techniques for demonstrating field biodegradation: measurements of field samples and experiments run in the field. In most bioremediation scenarios a third technique, modeling experiments, provides an improved understanding of the fate of contaminants in field sites. Because none of these three techniques alone can show with complete certainty that biodegradation is the primary cause of declining contaminant concentrations, the most effective strategy for demonstrating bioremediation usually combines several techniques.

Measurements of Field Samples

The following techniques for documenting in situ bioremediation involve analyzing the chemical and microbiological properties of soil and ground water samples from the contaminated site:

- **Number of bacteria**. Because microbes often reproduce when they degrade contaminants, an increase in the number of contaminant-degrading bacteria over usual conditions may indicate successful bioremediation.
- **Number of protozoans**. Because protozoans prey on bacteria, an increase in the number of protozoans signals bacterial population growth, indicating that bioremediation may be occurring.
- **Rates of bacterial activity**. Tests indicating that bacteria from the contaminated site degrade the contaminant rapidly enough to

effect remediation when incubated in microcosms that resemble the field site provide further evidence of successful bioremediation.

- **Adaptation**. Tests showing that bacteria from the bioremediation zone can metabolize the contaminant, while bacteria from outside the zone cannot (or do so more slowly), show that the bacteria have adapted to the contaminant and indicate that bioremediation may have commenced.
- **Carbon isotopes**. Isotopic ratios of the inorganic carbon (carbon dioxide, carbonate ion, and related compounds) from a soil or water sample showing that the contaminant has been transformed to inorganic carbon are a strong indicator of successful bioremediation.
- **Metabolic byproducts**. Tests showing an increase in the concentrations of known byproducts of microbial activity, such as carbon dioxide, provide a sign of bioremediation.
- **Intermediary metabolites**. The presence of metabolic intermediates—simpler but incompletely degraded forms of the contaminant—in samples of soil or water signals the occurrence of biodegradation.
- **Growth-stimulating materials**. A depletion in the concentration of growth-stimulating materials, such as oxygen, is a sign that microbes are active and may indicate bioremediation.
- **Ratio of nondegradable to degradable compounds**. An increase in the ratio of compounds that are difficult to degrade to those that are easily degraded indicates that bioremediation may be occurring.

Experiments Run in the Field

The following methods for evaluating whether microorganisms are actively degrading the contaminant involve conducting experiments in the field:

- **Stimulating bacteria within subsites**. When growth-stimulating materials such as oxygen and nutrients are added to one subsite within the contaminated area but not another, the relative rate of contaminant loss should increase in the stimulant-amended subsite. The contrast in contaminant loss between enhanced and unenhanced subsites can be attributed to bioremediation.
- **Measuring the stimulant uptake rate**. Growth-stimulating materials, such as oxygen, can be added to the site in pulses to determine the rate at which they are consumed. Relatively rapid loss of oxygen or other stimulants in the contaminated area compared to an uncontaminated area suggests successful bioremediation.

- **Monitoring conservative tracers.** Tracer compounds that are not biologically reactive can be added to the site to determine how much contaminant (or growth-stimulating material) is disappearing through nonbiological pathways and how much is being consumed by microorganisms.
- **Labeling contaminants.** Contaminants can be labeled with chemical elements that appear in metabolic end products when the contaminants are degraded, providing another mechanism for determining whether biodegradation is responsible for a contaminant's disappearance.

Modeling Experiments

A final set of techniques for evaluating whether bioremediation is occurring in the field uses models—sets of mathematical equations that quantify the contaminant's fate. Modeling techniques provide a framework for formally deciding what is known about contaminant behavior at field sites. When modelers have a high degree of confidence that the model accurately represents conditions at the site, modeling experiments can be used to demonstrate field biodegradation.

There are two general strategies for using models to evaluate bioremediation. The first strategy, useful when biodegradation is the main phenomenon controlling the contaminant's fate, is to model the abiotic processes to determine how much contaminant loss they account for. Bioremediation is indicated when the concentrations of contaminant actually found in field sites are significantly lower than would be expected from predictions based on abiotic processes (such as dilution, transport, and volatilization). The second strategy involves directly modeling the microbial processes to estimate the biodegradation rates. Direct modeling, while the intellectually superior approach, requires quantitative information about the detailed interactions between microbial populations and site characteristics. Because this information may be difficult to obtain, direct modeling is primarily a topic of academic research and is seldom a routinely applied procedure.

Four different types of models have been developed:

- **Saturated flow models.** These models describe where and how fast the water and dissolved contaminants flow through the saturated zone.
- **Multiphase flow models.** These models characterize the situation in which two or more fluids, such as water and a nonaqueous-phase contaminant or water and air, exist together in the subsurface.

- **Geochemical models**. These models analyze how a contaminant's chemical speciation is controlled by the thermodynamics of the many chemical and physical reactions that may occur in the subsurface.
- **Biological reaction rate models**. These models represent how quickly the microorganisms transform contaminants.

Because so many complex processes interact in the subsurface, ultimately two or more types of models may be required for a complete evaluation.

Limitations Inherent in Evaluating In Situ Bioremediation

Although microorganisms grown in the laboratory can destroy most organic contaminants, the physical realities of the subsurface—the low fluid flow rates, physical heterogeneities, unknown amounts and locations of contaminants, and the contaminants' unavailability to the microorganisms—make in situ bioremediation a technological challenge that carries inherent uncertainties. Three strategies can help minimize these uncertainties: (1) increasing the number of samples used to document bioremediation, (2) using models so that important variables are properly weighted and variables with little influence are eliminated, and (3) compensating for uncertainties by building safety factors and flexibility into the design of engineering systems. These strategies should play important roles in evaluating bioremediation projects.

While uncertainties should be minimized, it is important to recognize that no strategy can entirely eliminate the uncertainties, even for the best-designed systems. Given today's knowledge base, it is not possible to fully understand every detail of whether and how bioremediation is occurring. The goal in evaluating in situ bioremediation is to assess whether the weight of evidence from tests such as those described above makes a convincing case for successful bioremediation.

CONCLUSIONS: FUTURE PROSPECTS FOR BIOREMEDIATION

Bioremediation integrates the tools of many disciplines. As each of the disciplines advances and as new cleanup needs arise, opportunities for new bioremediation techniques will emerge. As these new techniques are brought into commercial practice, the importance of sound methods for evaluating bioremediation will increase.

The fundamental knowledge base underlying bioremediation is sufficient to begin implementing the three-part evaluation strategy the committee has recommended. However, further research and better education of those involved in bioremediation will improve the ability to apply the strategy and understanding of the fundamentals behind bioremediation.

Recommended Steps in Research

The committee recommends research in the following areas to improve evaluations of bioremediation:

- **Evaluation protocols**. Protocols for putting the three-part evaluation strategy into practice need to be developed and field tested through coordinated efforts involving government, industry, and academia.
- **Innovative site characterization techniques**. Rapid, reliable, and inexpensive site characterization techniques would simplify many of the evaluation techniques this report describes. Examples of relevant site measurements include distribution of hydraulic conductivities, contaminant concentrations associated with solid or other nonaqueous phases, native biodegradation potential, and abundance of different microbial populations.
- **Improved models**. Improvements in mathematical models would increase the ability to link chemical, physical, and biological phenomena occurring in the subsurface and to quantify how much contaminant loss occurs because of biodegradation.

Recommended Steps in Education

Steps need to be taken to improve the understanding of what bioremediation is and what it can and cannot do. The committee recommends three types of educational steps:

- **Training courses that selectively extend the knowledge bases of the technical personnel currently dealing with the uses or potential uses of in situ bioremediation**. This step explicitly recognizes that practitioners and regulators who already are dealing with complicated applications of bioremediation need immediate education about technical areas outside their normal expertise.
- **Formal education programs that integrate the principles and practices for the next generation of technical personnel**. This step explicitly recognizes the need to educate a new generation of technical personnel who have far more interdisciplinary training than is currently available in most programs.

- **Means for effective transfer of information among the different stakeholders involved in a project.** Effective transfer requires that all types of stakeholders participate, that all are invested in achieving a common product (such as a design, a report, or an evaluation procedure), and that sufficient time is allocated for sharing perceptions and achieving the product. This step may involve more time and more intensive interactions than have been the norm in the past.

In summary, in situ bioremediation is a technology whose full potential has not been realized. As the limitations of conventional ground water and soil cleanup technologies become more apparent, research into alternative cleanup technologies will intensify. Bioremediation is an especially attractive alternative because it is potentially less costly than conventional cleanup methods, it shows promise for reaching cleanup goals more quickly than pump-and-treat methods, and it results in less transfer of contaminants to other media. However, bioremediation presents a unique technological challenge. The combination of the intricacies of microbial processes and the physical challenge of monitoring both microorganisms and contaminants in the subsurface makes bioremediation difficult to understand, and it makes some regulators and clients hesitant to trust bioremediation as an appropriate cleanup strategy. The inherent complexity involved in performing bioremediation in situ means that special attention must be given to evaluating the success of a project. Whether a bioremediation project is intrinsic or engineered, the importance of a sound strategy for evaluating bioremediation will increase in the future as the search for improved cleanup technologies accelerates.

1

Introduction

In the past decade the United States has spent billions of dollars trying to clean up contaminated ground water and soils, the legacy of an era in which industry grew faster than knowledge about safe chemical disposal. Despite the large financial investment, ground water cleanup efforts are falling short of public expectations. Recent studies have revealed that, while conventional cleanup technologies have prevented the contamination problem from spreading, in most cases they are incapable of restoring the water to meet health-based standards in a reasonable time frame. Soil cleanups have been more successful in meeting regulatory standards. However, conventional soil cleanup methods may transfer contaminants to the air, posing risks that are not always acceptable to residents near the contaminated site. The limitations of conventional ground water cleanup technologies and the hazards of conventional soil cleanup methods have spurred investigations into in situ bioremediation, which uses microorganisms to destroy or immobilize contaminants in place. Bioremediation is a promising alternative to conventional cleanup technologies for both ground water and soil because it may be faster, safer, and less costly.

Conventional methods for ground water cleanup rely on pumping water to the surface and treating it there. Such pump-and-treat methods are slow; they require the withdrawal of large volumes of water to flush contaminants from aquifer solids, and they may leave behind reservoirs of contaminants that are lighter or denser than water and/or have low solubilities. By treating the problem close to its source, in situ, bioremediation speeds contaminant desorption and dissolution. Consequently, a cleanup that might require decades using pump-and-treat methods could possibly be completed in a few years with bioremediation. In addition, pump-and-treat methods do not destroy contaminants but simply bring them to the surface for treatment or disposal elsewhere. In situ bioremediation, on the other hand, can transform contaminants to harmless byproducts such as carbon dioxide and water.

Conventional methods for soil cleanup require digging up the contaminated soil and either incinerating it or burying it at a specially designed disposal site. Soil excavation and incineration may increase the exposure to contaminants for both the workers at the site and nearby residents. Furthermore, excavation and final disposal are extremely costly. By treating the soil in place, bioremediation reduces both the exposure risk and the cleanup cost.

Because bioremediation shows promise as an alternative to conventional environmental cleanup technologies, the number of vendors selling bioremediation has increased dramatically in recent years. Bioremediation is one of the fastest-growing sectors of the U.S. hazardous waste market. It is expected to become a $500 million per year industry by the year 2000. Yet despite the rapid growth in the use of this technology, bioremediation is not universally understood or trusted by those who must approve its use, and its success is a hotly debated issue.

A primary reason for the lack of understanding and mistrust of bioremediation is that the technology requires knowledge not only of such fields as environmental engineering and hydrology, which are important in conventional cleanup methods, but also of the complex workings of microorganisms. The potential for large profits, when combined with the mysteriousness of applying microorganisms, makes bioremediation attractive for "snake oil salesmen" who claim to be able to solve all types of contamination problems. Many buyers of cleanup services and regulators who approve cleanup plans lack the necessary background to evaluate whether a bioremediation project has a feasible design. Furthermore, they may be unsure how to evaluate whether an ongoing bioremediation project is progressing toward successful completion. Consequently, some regulators and clients approach bioremediation with skepticism, opting for more conventional technologies even when bioremediation is the most appropriate technology for a particular site.

The multidisciplinary nature of bioremediation presents problems not only for clients and regulators but also for the vendors of environmental cleanup services. These vendors face a challenge in integrating the wide range of disciplines needed to carry out bioremediation in the field. Communications among engineers, microbiologists, hydrologists, and chemists are complicated by each discipline's highly specialized concepts, tools, and jargon.

Even when all parties are knowledgeable and competent, evaluating the success of bioremediation (i.e., whether or not it is working) can cause confusion. Part of the confusion comes from the inherent complexity of the sites. Knowing with certainty the location and fate of all contaminants is impossible. However, evaluating the success of in situ bioremediation is further complicated by the multiple definitions of success set forth by those involved with the cleanup:

- Regulators want cleanup standards to be met.
- Clients want the cleanup to be cost effective.
- Bioremediation vendors and researchers want evidence that the cleanup was caused by microbial action—that the contaminant did not, for example, simply evaporate or migrate off site.

This report is designed for bioremediation clients, regulators, and vendors, who need to agree on how to define success appropriately. The report emphasizes ways to show that microorganisms aided cleanup efforts because the use of microorganisms is what distinguishes bioremediation from other technologies. Chapter 2 explains the principles of bioremediation, describing the microbiological processes that can be employed in bioremediation and providing practical guidance on what types of contaminants and site conditions are most amenable to bioremediation. Chapter 3 reviews the current state of the art in in situ bioremediation systems. These chapters provide the basis for understanding when in situ bioremediation is likely to work. Chapter 4—the most critical part of the report—presents methods and strategies for evaluating a bioremediation project in the testing or implementation phase. Finally, Chapter 5 suggests innovations that may improve the technology's capabilities in the future.

This report represents the opinions of the National Research Council's Committee on In Situ Bioremediation. The National Research Council established this committee in June 1992 and assigned it the specific task of preparing guidelines for evaluating whether an in situ bioremediation project, either proposed or in the implementation stage, is likely to reach cleanup goals. The committee includes representatives of all groups with an interest in bioremediation: buyers of bioremediation services, bioremediation contractors, environmental regulators, and academic researchers. The committee developed the framework for this report and the guidelines it presents at a one-week workshop in October 1992. Also included in this volume are seven background papers authored by committee members to represent the range of perspectives from which bioremediation can be viewed.

2

Principles of Bioremediation

The key players in bioremediation are bacteria—microscopic organisms that live virtually everywhere. Microorganisms are ideally suited to the task of contaminant destruction because they possess enzymes that allow them to use environmental contaminants as food and because they are so small that they are able to contact contaminants easily. In situ bioremediation can be regarded as an extension of the purpose that microorganisms have served in nature for billions of years: the breakdown of complex human, animal, and plant wastes so that life can continue from one generation to the next. Without the activity of microorganisms, the earth would literally be buried in wastes, and the nutrients necessary for the continuation of life would be locked up in detritus.

Whether microorganisms will be successful in destroying manmade contaminants in the subsurface depends on three factors: the type of organisms, the type of contaminant, and the geological and chemical conditions at the contaminated site. This chapter explains how these three factors influence the outcome of a subsurface bioremediation project. It reviews how microorganisms destroy contaminants and what types of organisms play a role in in situ bioremediation. Then, it evaluates which contaminants are most susceptible to bioremediation in the subsurface and describes the types of sites where bioremediation is most likely to succeed.

THE ROLE OF MICROBES IN BIOREMEDIATION

The goal in bioremediation is to stimulate microorganisms with nutrients and other chemicals that will enable them to destroy the contaminants. The bioremediation systems in operation today rely on microorganisms native to the contaminated sites, encouraging them to work by supplying them with the optimum levels of nutrients and other chemicals essential for their metabolism. Thus, today's bioremediation systems are limited by the capabilities of the native microbes. However, researchers are currently investigating ways to augment contaminated sites with nonnative microbes—including genetically engineered microorganisms—specially suited to degrading the contaminants of concern at particular sites. It is possible that this process, known as bioaugmentation, could expand the range of possibilities for future bioremediation systems.

Regardless of whether the microbes are native or newly introduced to the site, an understanding of how they destroy contaminants is critical to understanding bioremediation. The types of microbial processes that will be employed in the cleanup dictate what nutritional supplements the bioremediation system must supply. Furthermore, the byproducts of microbial processes can provide indicators that the bioremediation is successful.

How Microbes Destroy Contaminants

Although bioremediation currently is used commercially to clean up a limited range of contaminants—mostly hydrocarbons found in gasoline—microorganisms have the capability to biodegrade almost all organic contaminants and many inorganic contaminants. A tremendous variety of microbial processes potentially can be exploited, extending bioremediation's utility far beyond its use today. Whether the application is conventional or novel by today's standards, the same principles must be applied to stimulate the right type and amount of microbial activity.

Basics of Microbial Metabolism

Microbial transformation of organic contaminants normally occurs because the organisms can use the contaminants for their own growth and reproduction. Organic contaminants serve two purposes for the organisms: they provide a source of carbon, which is one of the basic building blocks of new cell constituents, and they provide electrons, which the organisms can extract to obtain energy.

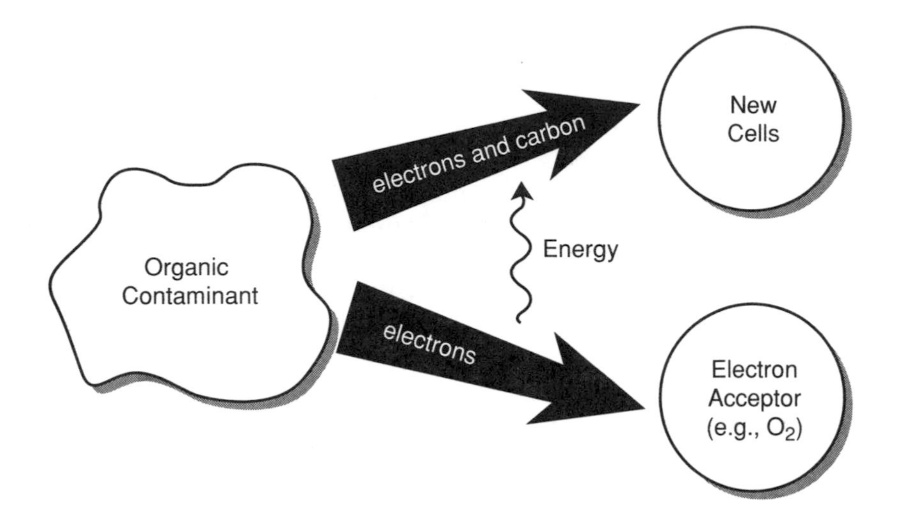

FIGURE 2-1 Microbes degrade contaminants because in the process they gain energy that allows them to grow and reproduce. Microbes get energy from the contaminants by breaking chemical bonds and transferring electrons from the contaminants to an electron acceptor, such as oxygen. They "invest" the energy, along with some electrons and carbon from the contaminant, to produce more cells.

Microorganisms gain energy by catalyzing energy-producing chemical reactions that involve breaking chemical bonds and transferring electrons away from the contaminant. The type of chemical reaction is called an *oxidation-reduction reaction*: the organic contaminant is *oxidized*, the technical term for losing electrons; correspondingly, the chemical that gains the electrons is *reduced*. The contaminant is called the *electron donor*, while the electron recipient is called the *electron acceptor*. The energy gained from these electron transfers is then "invested," along with some electrons and carbon from the contaminant, to produce more cells (see Figure 2-1). These two materials—the electron donor and acceptor—are essential for cell growth and are commonly called the *primary substrates*. (See Box 2-1 and the glossary for definitions of these and other key terms.)

Many microorganisms, like humans, use molecular oxygen (O_2) as the electron acceptor. The process of destroying organic compounds with the aid of O_2 is called *aerobic respiration*. In aerobic respiration, microbes use O_2 to oxidize part of the carbon in the contaminant to carbon dioxide (CO_2), with the rest of the carbon used to produce new cell mass. In the process the O_2 gets reduced, produc-

BOX 2-1
KEY TERMS FOR UNDERSTANDING BIOREMEDIATION

Microorganism: An organism of microscopic size that is capable of growth and reproduction through biodegradation of "food sources," which can include hazardous contaminants.

Microbe: The shortened term for microorganism.

Oxidize: The transfer of electrons away from a compound, such as an organic contaminant. The coupling of oxidation to reduction (see below) usually supplies energy that microorganisms use for growth and reproduction. Often (but not always), oxidation results in the addition of an oxygen atom and/or the loss of a hydrogen atom.

Reduce: The transfer of electrons to a compound, such as oxygen, that occurs when another compound is oxidized.

Electron acceptor: The compound that receives electrons (and therefore is reduced) in the energy-producing oxidation-reduction reactions that are essential for the growth of microorganisms and bioremediation. Common electron acceptors in bioremediation are oxygen, nitrate, sulfate, and iron.

Electron donor: The compound that donates electrons (and therefore is oxidized). In bioremediation the organic contaminant often serves as an electron donor.

Primary substrates: The electron donor and electron acceptor that are essential to ensure the growth of microorganisms. These compounds can be viewed as analogous to the food and oxygen that are required for human growth and reproduction.

Aerobic respiration: The process whereby microorganisms use oxygen as an electron acceptor.

Anaerobic respiration: The process whereby microorganisms use a chemical other than oxygen as an electron acceptor. Common "substitutes" for oxygen are nitrate, sulfate, and iron.

Fermentation: The process whereby microorganisms use an organic compound as both electron donor and electron acceptor, converting the compound to fermentation products such as organic acids, alcohols, hydrogen, and carbon dioxide.

continued

Cometabolism: A variation on biodegradation in which microbes transform a contaminant even though the contaminant cannot serve as the primary energy source for the organisms. To degrade the contaminant, the microbes require the presence of other compounds (primary substrates) that can support their growth.

Reductive dehalogenation: A variation on biodegradation in which microbially catalyzed reactions cause the replacement of a halogen atom on an organic compound with a hydrogen atom. The reactions result in the net addition of two electrons to the organic compound.

Intrinsic bioremediation: A type of bioremediation that manages the innate capabilities of naturally occurring microbes to degrade contaminants without taking any engineering steps to enhance the process.

Engineered bioremediation: A type of remediation that increases the growth and degradative activity of microorganisms by using engineered systems that supply nutrients, electron acceptors, and/or other growth-stimulating materials.

ing water. Thus, the major byproducts of aerobic respiration are carbon dioxide, water, and an increased population of microorganisms.

Variations on Basic Metabolism

In addition to microbes that transform contaminants through aerobic respiration, organisms that use variations on this basic process have evolved over time. These variations allow the organisms to thrive in unusual environments, such as the underground, and to degrade compounds that are toxic or not beneficial to other organisms.

Anaerobic Respiration. Many microorganisms can exist without oxygen, using a process called *anaerobic respiration*. In anaerobic respiration, nitrate (NO_3^-), sulfate (SO_4^{2-}), metals such as iron (Fe^{3+}) and manganese (Mn^{4+}), or even CO_2 can play the role of oxygen, accepting electrons from the degraded contaminant. Thus, anaerobic respiration uses inorganic chemicals as electron acceptors. In addition to new cell matter, the byproducts of anaerobic respiration may include nitrogen gas (N_2), hydrogen sulfide (H_2S), reduced forms of metals, and methane (CH_4), depending on the electron acceptor.

Some of the metals that anaerobic organisms use as electron acceptors are considered contaminants. For example, recent research has demonstrated that some microorganisms can use soluble uranium (U^{6+}) as an electron acceptor, reducing it to insoluble uranium (U^{4+}). Under this circumstance the organisms cause the uranium to precipitate, decreasing its concentration and mobility in the ground water.

Inorganic Compounds as Electron Donors. In addition to organisms that use inorganic chemicals as electron acceptors for anaerobic respiration, other organisms can use inorganic molecules as electron donors. Examples of inorganic electron donors are ammonium (NH_4^+), nitrite (NO_2^-), reduced iron (Fe^{2+}), reduced manganese (Mn^{2+}), and H_2S. When these inorganic molecules are oxidized (for example, to NO_2^-, NO_3^-, Fe^{3+}, Mn^{4+}, and SO_4^{2-}, respectively), the electrons are transferred to an electron acceptor (usually O_2) to generate energy for cell synthesis. In most cases, microorganisms whose primary electron donor is an inorganic molecule must obtain their carbon from atmospheric CO_2 (a process called CO_2 fixation).

Fermentation. A type of metabolism that can play an important role in oxygen-free environments is *fermentation.* Fermentation requires no external electron acceptors because the organic contaminant serves as both electron donor *and* electron acceptor. Through a series of internal electron transfers catalyzed by the microorganisms, the organic contaminant is converted to innocuous compounds known as fermentation products. Examples of fermentation products are acetate, propionate, ethanol, hydrogen, and carbon dioxide. Fermentation products can be biodegraded by other species of bacteria, ultimately converting them to carbon dioxide, methane, and water.

Secondary Utilization and Cometabolism. In some cases, microorganisms can transform contaminants, even though the transformation reaction yields little or no benefit to the cell. The general term for such nonbeneficial biotransformations is *secondary utilization*, and an important special case is called *cometabolism.* In cometabolism the transformation of the contaminant is an incidental reaction catalyzed by enzymes involved in normal cell metabolism or special detoxification reactions. For example, in the process of oxidizing methane, some bacteria can fortuitously degrade chlorinated solvents that they would otherwise be unable to attack. When the microbes oxidize methane, they produce certain enzymes that incidentally destroy the chlorinated solvent, even though the solvent itself cannot support microbial growth. The methane is the primary electron donor because it is the organisms' primary food source, while the chlorinated solvent is a *secondary substrate* because it does not support the bacteria's growth. In addition to methane, toluene and phenol have been used as primary substrates to stimulate cometabolism of chlorinated solvents.

Reductive Dehalogenation. Another variation on microbial metabolism is *reductive dehalogenation.* Reductive dehalogenation is potentially important in the detoxification of halogenated organic contaminants, such as chlorinated solvents. In reductive dehalogenation, microbes catalyze a reaction in which a halogen atom (such as chlorine) on the contaminant molecule gets replaced with a hydrogen atom. The reaction adds two electrons to the contaminant molecule, thus reducing the contaminant.

For reductive dehalogenation to proceed, a substance other than the halogenated contaminant must be present to serve as the electron donor. Possible electron donors are hydrogen and low-molecular-weight organic compounds (lactate, acetate, methanol, or glucose). In most cases, reductive dehalogenation generates no energy but is an incidental reaction that may benefit the cell by eliminating a toxic material. However, researchers are beginning to find examples in which cells can obtain energy from this metabolic process.

Microbial Nutritional Requirements for Contaminant Destruction

Regardless of the mechanism microbes use to degrade contaminants, the microbes' cellular components have relatively fixed elemental compositions. A typical bacterial cell is 50 percent carbon; 14 percent nitrogen; 3 percent phosphorus; 2 percent potassium; 1 percent sulfur; 0.2 percent iron; and 0.5 percent each of calcium, magnesium, and chloride. If any of these or other elements essential to cell building is in short supply relative to the carbon present as organic contaminants, competition for nutrients within the microbial communi ties may limit overall microbial growth and slow contaminant removal. Thus, the bioremediation system must be designed to supply the proper concentrations and ratios of these nutrients if the natural habitat does not provide them.

How Microbes Demobilize Contaminants

In addition to converting contaminants to less harmful products, microbes can cause mobile contaminants to be demobilized, a strategy useful for containing hazardous materials. There are three basic ways microbes can be used to demobilize contaminants:

- Microbial biomass can sorb hydrophobic organic molecules. Sufficient biomass grown in the path of contaminant migration could stop or slow contaminant movement. This concept is sometimes called a biocurtain.
- Microorganisms can produce reduced or oxidized species that cause metals to precipitate. Examples are oxidation of Fe^{2+} to Fe^{3+}, which precipitates as ferric hydroxide ($FeOH_{3(s)}$); reduction of SO_4^{2-} to sulfide (S^{2-}), which precipitates with Fe^{2+} as pyrite ($FeS_{(s)}$) or with mercury (Hg^{2+}) as mercuric sulfide ($HgS_{(s)}$); reduction of hexavalent chromium (Cr^{6+}) to trivalent chromium (Cr^{3+}), which can precipitate as chromium oxides, sulfides, or phosphates; and, as mentioned previously, reduction of soluble uranium to insoluble U^{4+}, which precipitates as uraninite (UO_2).
- Microorganisms can biodegrade organic compounds that bind with metals and keep the metals in solution. Unbound metals often precipitate and are immobilized.

Indicators of Microbial Activity

In the process of degrading or demobilizing contaminants, microbes cause changes in the surrounding environment that are important to understand when evaluating bioremediation.

Chemical Changes

Bioremediation alters the ground water chemistry. These chemical changes follow directly from the physiological principles of microorganisms outlined above. Microbial metabolism catalyzes reactions that consume well-defined reactants—contaminants and O_2 or other electron acceptors—converting them to well-defined products.

The specific chemical reactants and products can be determined from the chemical equations for the reactions the microbes catalyze. These equations are familiar to anyone with a basic understanding of microbiology. For example, the chemical equation for the degradation of toluene (C_7H_8) is:

$$C_7H_8 + 9O_2 ========> 7CO_2 + 4H_2O$$

Thus, when bioremediation is occurring, the concentration of inorganic carbon (represented by CO_2) should increase as the concentrations of toluene and oxygen decrease. Another example is the dechlorination of trichloroethane ($C_2H_3Cl_3$, or TCA) to dichloroethane ($C_2H_4Cl_2$, or DCA) by hydrogen-oxidizing anaerobic bacteria:

$$C_2H_3Cl_3 + H_2 ========> C_2H_4Cl_2 + H^+ + Cl^-$$

Here, TCA and hydrogen (H_2) decrease as DCA, hydrogen ion (H^+), and chloride ion (Cl^-) increase. The formation of hydrogen ion may cause the pH to decrease, depending on the ground water chemistry.

In general, under aerobic conditions, one should expect to observe a drop in the O_2 concentration when microbes are active. Similarly, under anaerobic conditions, concentrations of other electron acceptors—NO_3^-, SO_4^{2-}, Fe^{3+}, Mn^{4+}—will decrease, with a corresponding increase in the reduced species of these compounds (N_2, H_2S, Fe^{2+}, and Mn^{2+}, respectively). Under both types of conditions the inorganic carbon concentration should increase, because organic carbon is oxidized. The inorganic carbon may take the form of gaseous CO_2, dissolved CO_2, or bicarbonate ion (HCO_3^-).

Adaptation by Native Organisms

In addition to producing chemical changes in the ground water, bioremediation can alter the metabolic capabilities of native microorganisms. Often, microorganisms do not degrade contaminants upon initial exposure, but they may develop the capability to degrade the contaminant after prolonged exposure. Several mechanisms have been proposed to explain metabolic adaptation, including enzyme induction, growth of biodegrading populations, and genetic change. However, these proposals remain largely speculative because methodological limitations usually preclude rigorous understanding of how microbial communities develop, both in laboratory tests and at field sites. Regardless of the mechanisms, adaptation is important because it is a critical principle in ensuring the existence of microorganisms that can destroy the myriad new chemicals that humans have created and introduced into the environment.

Adaptation occurs not only within single microbial communities but also among distinct microbial communities that may evolve a cooperative relationship in the destruction of compounds. One community may partially degrade the contaminant, and a second community farther along the ground water flow path may complete the reaction. This type of coupling occurs naturally in anaerobic food chains that convert plant-derived organic compounds to methane. Such coupling has obvious applications for bioremediation of sites bearing contaminant compounds whose complete metabolism may require alternation between anaerobic and aerobic processes.

Growth of Predators

Although bacteria are the agents for biodegradation during bioremediation, other organisms that prey on bacteria also may grow as a result of bioremediation. Protozoa are the most common bacterial predators. Just as mammalian predators, such as wolves, can only be supported by certain densities of their prey, microbial protozoan predators flourish only when their bacterial prey are in large, rapidly replenished supplies. Thus, the presence of protozoa normally signifies that enough bacteria have grown to degrade a significant quantity of contaminants.

Complicating Factors

The basic principles of how microbes degrade contaminants are relatively straightforward. Yet many details of microbial metabolism are not yet understood, and the successful use of microbes in bioremediation is not a simple matter. A range of factors may complicate bioremediation. Some of the key complicating factors are the unavailability of contaminants to the organisms, toxicity of contaminants to the organisms, microbial preference for some contaminants or naturally occurring chemicals over other contaminants, partial degradation of contaminants to produce hazardous byproducts, inability to remove contaminants to very low concentrations, and aquifer clogging from excessive biomass growth.

Unavailability of Contaminants to the Organisms

Readily biodegradable contaminants may remain undegraded or be biodegraded very slowly if their concentrations in the ground water are too low. The problem of too low concentrations usually is caused by unavailability, in which the contaminant is sequestered from the microorganisms. Sequestering of organic contaminants can occur when the contaminant is dissolved in a nonaqueous-phase liquid—a solution that does not mix easily with water and therefore travels through the ground separately from the ground water. Sequestering of organic contaminants can also occur if the contaminant is strongly adsorbed to soil surfaces or is trapped in pores too small for circulating ground water to penetrate easily. In these cases, almost all of the contaminant is associated with the solid, the nonaqueous-phase liquid, or the pores, and the very small concentrations that dissolve in the water support very small or zero biodegradation rates.

Sequestering of metals and other inorganic contaminants occurs most frequently when they precipitate.

One possible strategy for overcoming the unavailability problem is to add chemical agents that mobilize the contaminants, causing them to move with the ground water. Such chemical agents are already used at some sites to increase the efficiency of conventional pump-and-treat ground water cleanup systems. However, their use to facilitate bioremediation is more complex than their use for pump-and-treat systems because the mobilizing agents not only affect the physical properties of the contaminants but may also affect the activity of the microorganisms.

Organic contaminants can be mobilized by adding surfactants. When only small surfactant concentrations are applied, the surfactant molecules accumulate at solid surfaces, reduce the surface tension, and, in principle, increase the spreading of organic contaminants. This spreading might improve contaminant transfer to the water and thereby accelerate bioremediation, but evidence is not clear for actual subsurface conditions. When large concentrations of surfactant are added, the surfactant molecules join together in colloids, called micelles. Organic contaminants dissolve into the micelles and are transported with the water inside them. However, biodegradation usually is not enhanced by contaminant transfer into the micelles because the true aqueous-phase concentration is not increased.

Metals can be mobilized by adding chemicals called complexing agents, or ligands, to which the metals bond. The formation of metal-ligand bonds dissolves precipitated metals, increasing their mobility. However, the effectiveness of strong ligands, such as ethylenediaminetetraacetic acid (EDTA), in enhancing biodegradation is not yet proven. One potential limitation of using ligands to mobilize metals is that microbes may degrade the ligands, releasing the metals and causing them to precipitate again.

In some cases, bacteria produce their own surfactants and ligands that are useful in mobilizing trapped contaminants. In these cases the main purpose of the microorganisms is to produce mobilizing agents, not to biodegrade the contaminants. Bacterially mediated mobilization makes trapped contaminants more accessible for cleanup with pump-and-treat technology; it is potentially less costly than injecting commercial surfactants.

Toxicity of Contaminants to the Organisms

Just as contaminant concentrations that are too low can complicate bioremediation, high aqueous-phase concentrations of some contaminants can create problems. At high concentration, some chemicals are toxic to microbes, even if the same chemicals are readily biodegraded at lower concentrations. Toxicity prevents or slows metabolic reactions and often prevents the growth of the new biomass needed to stimulate rapid contaminant removal. The degree and mechanisms of toxicity vary with specific toxicants, their concentration, and the exposed microorganisms. Microbial cells cease to function when at least one of the essential steps in their myriad physiological processes is blocked. The blockage may result from gross physical disruption of cell structure or competitive binding of a single enzyme essential for metabolizing the toxicant.

Presence of Multiple Contaminants and Natural Organic Chemicals

Frequently, contaminated sites contain a combination of several man-made organic contaminants and naturally occurring organic chemicals from decayed plant and animal matter. When such mixtures of organics are present, microbes may selectively degrade the compound that is easiest to digest or that provides the most energy. Microbiologists have long been aware that complex mechanisms regulating microbial metabolism may cause some carbon compounds to be ignored while others are selectively used. This phenomenon, known as diauxy, could have serious implications for bioremediation efforts if the targeted contaminant is accompanied by substantial quantities of preferred growth substrates.

Mixtures do not always cause problems and sometimes can promote bioremediation. For example, biomass that grows primarily to degrade one type of organic compound may also degrade a second compound present at a concentration too low to support bacterial growth by itself.

Incomplete Degradation of Contaminants

In some cases, contaminants may not be fully degraded by the organisms. Partial degradation may diminish the concentration of the original pollutant but create metabolic intermediates that in some cases are more toxic than the parent compound. There are two main reasons why intermediates build up. In one case a so-called dead-end product is produced. Dead-end products may form during cometabolism, because the incidental metabolism of the contaminant may create a product that the bacterial enzymes cannot transform any further. For example, in the cometabolism of chlorinated phenols, dead-end products such as chlorocatechols, which are toxic, sometimes build up. In the second case the intermediate builds up even though the compound can be fully degraded, because some of the key bacterially mediated reactions are slow. For example, vinyl chloride, a cancer-causing agent, may build up during trichloroethylene (TCE) biodegradation. The bacteria can convert TCE to vinyl chloride relatively quickly, but the subsequent degradation of vinyl chloride usually occurs slowly.

Inability to Remove Contaminants to Low Concentrations

Microorganisms may sometimes be physiologically incapable, even when environmental conditions are optimal, of reducing pollutant concentrations to very low, health-based levels, because the uptake and metabolism of organic compounds sometimes stops at low concentrations. This may be caused by the cells' internal mechanisms for regulating what reactions they perform or by an inability of the capable microbial populations to survive given inadequate sustenance. Regardless of the mechanism, if the final contaminant concentration fails to meet the cleanup goal, other cleanup strategies (microbiological or other) may have to be implemented to effectively reduce the concentration to acceptable levels. Research on augmenting sites with nonnative microbes and controlling cells' genetic capabilities and internal regulation may lead to means for overcoming this limitation.

Aquifer Clogging

Stimulating the growth of enough microorganisms to ensure contaminant degradation is essential to in situ bioremediation. However, if all the organisms accumulate in one place, such as near the wells that supply growth-stimulating nutrients and electron acceptors, microbial growth can clog the aquifer. Clogging can interfere with effective circulation of the nutrient solution, limiting bioremediation in places that the solution does not reach. Protozoan predators may help mitigate clogging. In addition, two engineering strategies can help prevent clogging: (1) feeding nutrients and substrates in alternating pulses and (2) adding hydrogen peroxide as the oxygen source. Pulse feeding prevents excessive biomass growth by ensuring that high concentrations of all the growth-stimulating materials do not accumulate near the injection point. Hydrogen peroxide prevents excessive growth because it is a strong disinfectant, until it decomposes to oxygen and water.

CONTAMINANTS SUSCEPTIBLE TO BIOREMEDIATION

A critical factor in deciding whether bioremediation is the appropriate cleanup remedy for a site is whether the contaminants are susceptible to biodegradation by the organisms at the site (or by organisms that could be successfully grown at the site). Some compounds are more easily degraded by a wide range of organisms than others, and systems for encouraging biodegradation are better established for some compounds than for others. Table 2-1 provides an overview of classes of compounds and their inherent suitability for bioremediation. The table is intended to deliver a broad perspective on how chemical and microbiological properties jointly affect prospects for bioremediation, and the judgments it presents are generalities that, of course, have exceptions. The table shows that bioremediation treatment technology is well established for certain classes of petroleum hydrocarbons but that the technologies for treating all other classes are still emerging. Commercial development of bioremediation technologies for these other compounds is possible, but it will require further research and the scaling up of lab discoveries for application in the field.

The table's first column shows the contaminant's frequency of occurrence at hazardous waste sites. It indicates the magnitude of the problem the contaminant poses. The second column indicates the state of development of bioremediation technologies for cleaning up the contaminant. In this column, "established" means that bioremediation of the contaminant has been tried successfully many times at the commercial scale. "Emerging" means that the concepts underlying bioremediation of the contaminant have been tested in the laboratory and, in some cases, tested successfully at a limited number of field sites under controlled conditions. "Possible" means that evidence from lab tests indicates future potential for bioremediation to successfully clean up the compound. The third column presents the evidence leading the committee to believe that the contaminant can be cleaned up successfully with bioremediation in the future, even though established bioremediation technology does not yet exist. It indicates what types of organisms can degrade the contaminant and how easily they can act. The fourth column describes contaminant properties that may limit bioremediation. The key limiting properties are the contaminant's tendency to sorb to subsurface solids and to partition into a nonaqueous phase that travels separately from the ground water. As discussed previously in this chapter, both of these properties—sorption and nonaqueous-phase formation—decrease the amount of contaminant available to the microorganisms, slowing

TABLE 2-1 Contaminant Susceptibility to Bioremediation

Chemical Class	Frequency of Occurrence	Status of Bioremediation	Evidence of Future Success	Limitations
Hydrocarbons and Derivatives				
Gasoline, fuel oil	Very frequent	Established		Forms nonaqueous-phase liquid
Polycyclic aromatic hydrocarbons	Common	Emerging	Aerobically biodegradable under a narrow range of conditions	Sorbs strongly to subsurface solids
Creosote	Infrequent	Emerging	Readily biodegradable under aerobic conditions	Sorbs strongly to subsurface solids; forms nonaqueous-phase liquid
Alcohols, ketones, esters	Common	Established		
Ethers	Common	Emerging	Biodegradable under a narrow range of conditions using aerobic or nitrate-reducing microbes	
Halogenated Aliphatics				
Highly chlorinated	Very frequent	Emerging	Cometabolized by anaerobic microbes; cometabolized by aerobes in special cases	Forms nonaqueous-phase liquid
Less chlorinated	Very frequent	Emerging	Aerobically biodegradable under a narrow range of conditions; cometabolized by anaerobic microbes	Forms nonaqueous-phase liquid

Halogenated Aromatics				
Highly chlorinated	Common	Emerging	Aerobically biodegradable under a narrow range of conditions; cometabolized by anaerobic microbes	Sorbs strongly to subsurface solids; forrns nonaqueous phase—solid or liquid
Less chlorinated	Common	Emerging	Readily biodegradable under aerobic conditions	Forrns nonaqueous phase—solid or liquid
Polychlorinated Biphenyls				
Highly chlorinated	Infrequent	Emerging	Cometabolized by anaerobic microbes	Sorbs strongly to subsurface solids
Less chlorinated	Infrequent	Emerging	Aerobically biodegradable under a narrow range of conditions	Sorbs strongly to subsurface solids
Nitroaromatics	Common	Emerging	Aerobically biodegradable; converted to innocuous volatile organic acids under anaerobic conditions	
Metals (Cr, Cu, Ni, Pb, Hg, Cd, Zn, etc.)	Common	Possible	Solubility and reactivity can be changed by a variety of microbial processes	Availability highly variable—controlled by solution and solid chemistry

bioremediation. In general, the least biodegradable contaminants are those with the strongest tendency to sorb.

The table groups contaminants into five classes: petroleum hydrocarbons and derivatives, halogenated aliphatics, halogenated aromatics, nitroaromatics, and metals. Each class is discussed in more detail below.

Petroleum Hydrocarbons and Derivatives

Petroleum hydrocarbons and their derivatives are naturally occurring chemicals that humans have exploited for a wide range of purposes, from fueling engines to manufacturing chemicals. The representative types of petroleum hydrocarbons and derivatives listed in Table 2-1 are gasoline, fuel oil, polycyclic aromatic hydrocarbons (PAHs), creosote, ethers, alcohols, ketones, and esters. Each of these chemicals has a broad range of industrial applications. For example, PAHs are released when crude oil is refined and from the manufacture of petroleum products such as plastics. Creosote is used in wood preservatives. Ethers, esters, and ketones are components of chemicals ranging from perfumes, to anesthetics, to paints and lacquers, to insecticides.

Gasoline, fuel oil, alcohols, ketones, and esters have been successfully bioremediated at contaminated sites via established bioremediation procedures. Gasoline, in particular, has been the focus of substantial biodegradation and bioremediation research. The gasoline components benzene, toluene, ethylbenzene, and xylene (together known as BTEX) are relatively easy to bioremediate for several reasons:

- They are relatively soluble compared to other common contaminants and other gasoline components.
- They can serve as the primary electron donor for many bacteria widely distributed in nature.
- They are rapidly degraded relative to other contaminants shown in Table 2-1.
- The bacteria that degrade BTEX grow readily if oxygen is available.

Under many circumstances, ether bonds show considerable chemical stability and therefore resist microbial attack. High-molecular-weight compounds such as creosotes and some PAHs are also only slowly metabolized—partly as a result of their structural complexity, low solubility, and strong sorptive characteristics. Thus, bioremediation techniques for these latter classes of petroleum derivatives are still emerging.

Halogenated Compounds

Halogenated compounds are compounds with halogen atoms (usually chlorine, bromine, or fluorine) added to them in place of hydrogen atoms. Although halogenated organic compounds have been found in nature, these are not significant compared to the synthetic chemicals listed in the middle portion of Table 2-1. When halogen atoms are introduced into organic molecules, many properties, such as solubility, volatility, density, and toxicity, change markedly. These changes confer improvements that are valuable for commercial products, such as solvents used for degreasing, but they also have serious implications for microbial metabolism. The susceptibility of the chemicals to enzymatic attack is sometimes drastically decreased by halogenation, and persistent compounds often result. Consequently, bioremediation technologies for these chemicals are still emerging.

There are two broad classes of halogenated chemicals: halogenated aliphatics and halogenated aromatics.

Halogenated Aliphatics

Halogenated aliphatic compounds are compounds built from straight chains of carbon and hydrogen with varying numbers of hydrogen atoms replaced by halogen atoms. Halogenated aliphatics are effective solvents and degreasers and have been widely used in manufacturing and service industries, ranging from automobile manufacturing to dry cleaning. Some highly chlorinated representatives of this class, such as tetrachloroethene, are completely resistant to attack by aerobic microbes but are susceptible to degradation by special classes of anaerobic organisms. In fact, recent evidence shows that certain anaerobes can completely dechlorinate tetrachloroethene to the relatively nontoxic compound ethene, which is readily decomposed by aerobic microbes.

As the degree of halogenation in aliphatics diminishes, susceptibility to aerobic metabolism increases. The less halogenated ethenes may be destroyed by cometabolism when certain aerobic microbes are supplied with methane, toluene, or phenol, as described earlier in this chapter. Thus, a common treatment rationale for the highly chlorinated aliphatics is to remove the chlorine atoms anaerobically, with methanogens, and then complete the biodegradation process using aerobic cometabolism. However, routine procedures for implementing anaerobic/aerobic sequencing to bioremediate sites contaminated with chlorinated aliphatic materials are not yet established at the commercial scale.

Halogenated Aromatics

Halogenated aromatics are compounds built from one or more halogen-bearing benzene rings. Examples include chlorinated benzenes, used as solvents and pesticides; pentachlorophenol, used in fungicides and herbicides; and polychlorinated biphenyls (PCBs), once widely used in electrical transformers and capacitors. The aromatic benzene nucleus is susceptible to aerobic and anaerobic metabolism, although the latter occurs relatively slowly. Overall, however, the presence of halogen atoms on the aromatic ring governs biodegradability. A high degree of halogenation may prevent aromatic compounds from being aerobically metabolized, as is the case for highly chlorinated PCBs. However, as discussed above for the aliphatic compounds, anaerobic microbes can remove chlorine atoms from the highly halogenated aromatics. As the halogen atoms are replaced by hydrogen atoms, the molecules become susceptible to aerobic attack. Thus, a possible bioremediation scenario for treating soils, sediments, or water contaminated with halogenated aromatic chemicals is anaerobic dehalogenation followed by aerobic destruction of the residual compounds. It should be noted, however, that when certain substituent groups accompany the halogens on the aromatic ring, aerobic metabolism may proceed rapidly, as is the case for pentachlorophenol.

Nitroaromatics

Nitroaromatics are organic chemicals in which the nitro group ($-NO_2$) is bonded to one or more carbons in a benzene ring. A familiar example is trinitrotoluene (TNT), which is used in explosives. Laboratory research has shown that both aerobic and anaerobic microbes can convert many of these compounds to carbon dioxide, water, and mineral components. Recent field tests have confirmed that anaerobic microbes can transform nitroaromatics to innocuous volatile organic acids, like acetate, which then may be mineralized.

Metals

The metals listed in Table 2-1 are common pollutants inadvertently released during the manufacture of various industrial products, from steel to pharmaceuticals. Microorganisms cannot destroy metals, but they can alter their reactivity and mobility. Schemes for using microorganisms to mobilize metals from one location and scavenge the metal from another location have been applied to mining operations. Microbes produce acids that can leach metals, like copper, from low-grade ores. This same approach should be feasible for bioremediation purposes, but it has not been proven. Microorganisms can also demobilize metals by transforming them to a form that precipitates (see "How Microbes Demobilize Contaminants," earlier in this chapter).

ENVIRONMENTS AMENABLE TO BIOREMEDIATION

The suitability of a site for bioremediation depends not only on the contaminant's biodegradability but also on the site's geological and chemical characteristics. The ideal site for in situ bioremediation is one that is as controllable and easy to interpret as the small, laboratory-incubated flask experiments used to test pollutant biodegradation. The site most amenable to bioremediation, like the lab flask, has favorable chemical characteristics and relatively uniform geology. Site characteristics are rarely ideal, however. Each site is a unique section of landscape that presents an unpredictable variety of environmental conditions. Properties such as soil type, geological strata, and water chemistry vary not only from site to site but also within an individual site. Furthermore, site complexity and lack of site data commonly obscure the true type and severity of the contamination. It is normal in implementing bioremediation—or any other cleanup technology—to revise cleanup plans continually as more information becomes available during the remediation.

It is important to realize that no single set of site characteristics will favor bioremediation of all chemical contaminants. For example, certain compounds can only be metabolized under anaerobic conditions, while metabolism of others requires oxygen. When the degradation mechanisms for two co-occurring contaminants are mutually exclusive, difficult choices need to be made or sequential treatment schemes need to be devised.

Two Types of Bioremediation: Intrinsic and Engineered

A principal concern in determining whether the site environment is appropriate for in situ bioremediation is the type of bioremediation to be implemented. Bioremediation can be grouped into two broad types: *intrinsic* and *engineered*. Figure 2-2 illustrates the differences between the two.

Intrinsic bioremediation manages the innate capabilities of naturally occurring microbial communities to degrade environmental pollutants without taking any engineering steps to enhance the process. It differs from no-action alternatives in that it requires thorough docu-

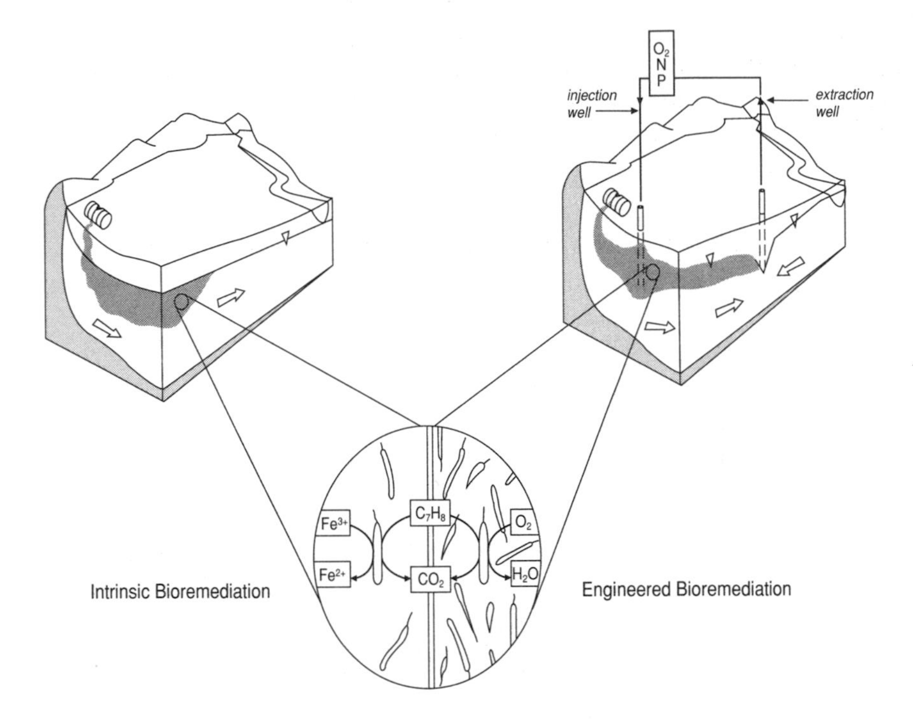

FIGURE 2-2 The differences between intrinsic and engineered bioremediation. In intrinsic bioremediation, *left*, native subsurface microbes degrade the contaminants without direct human intervention. In the close-up view, the native microbes use iron (Fe^{3+}) as an electron acceptor to degrade toluene (C_7H_8), a representative contaminant, and convert it to carbon dioxide (CO_2). In engineered bioremediation, *right*, oxygen (O_2), nitrogen (N), and phosphorus (P) are circulated through the subsurface via an injection and extraction well to promote microbial growth. In this case the microbes use oxygen as the electron acceptor, converting it to water (H_2O) as they degrade the toluene. Note that, as pictured in the close-up view, considerably more microbes are present in the engineered bioremediation system than in the intrinsic system. Consequently, contaminant degradation occurs more quickly in the engineered system. Intrinsic bioremediation requires extensive monitoring to ensure that the contaminant does not advance more quickly than the native microbes can degrade it.

mentation of the role of native microorganisms in eliminating contaminants via tests performed at field sites or on site-derived samples of soil, sediment, or water. Furthermore, the effectiveness of intrinsic bioremediation must be proven with a site-monitoring regime that routinely analyzes contaminant concentrations. The terms "natural," "passive," and "spontaneous" bioremediation and "bioattenuation"

BOX 2-2
INTRINSIC BIOREMEDIATION OF A CRUDE OIL SPILL—BEMIDJI, MINNESOTA

In August 1979 an oil pipeline burst near Bemidji, Minnesota, spilling approximately 100,000 gallons of crude oil into the surrounding ground water and soil. In 1983 researchers from the U.S. Geological Survey (USGS) began monitoring the site carefully to determine the crude oil's fate. They discovered that, although components of the crude oil initially migrated a short distance, native microorganisms capable of degrading the oil have prevented widespread contamination of the ground water. The microbes went to work with no human intervention, showing that intrinsic bioremediation can be effective for containing spills of petroleum products.

In the years following the spill, portions of the crude oil dissolved in the flowing ground water and moved 200 m from the original spill site. The undissolved crude oil itself migrated 30 m in the direction of ground water flow, and crude oil vapors moved 100 m in the overlying soil. However, the USGS researchers' detailed monitoring shows that the contaminant plume has not advanced since 1987, and the researchers have attributed this halt to intrinsic bioremediation.

Three types of evidence convinced the researchers that intrinsic bioremediation was largely responsible for containing the crude oil. First, modeling studies showed that if the oil were not biodegradable, the plume would have spread 500 to 1200 m since the spill (see Figure 2-3). Second, the concentrations of Fe^{2+} and CH_4 increased dramatically in the portion of the contaminant plume where oxygen was not present—evidence of an increase in activity by anaerobic organisms capable of degrading certain crude oil components, such as toluene. Third, concentrations of the crude oil components benzene and ethylbenzene, which are susceptible to aerobic degradation but less susceptible to anaerobic degradation, remained relatively stable in the anaerobic portion of the plume but decreased dramatically at the outer edges of the plume, where mixing with oxygenated water allowed aerobic degradation to occur.

The evidence from this site shows that, in hydrologic settings where intrinsic bioremediation rates are fast relative to hydrologic transport rates, native microbes can effectively confine contaminants to near the spill source without further human intervention. However, it is essential for such sites to have detailed, long-term monitoring plans to ensure that the contamination is, indeed, contained. At some sites, the rates of hydrologic transport outpace the rates of intrinsic bioremediation, and additional engineering steps to contain or remove the contamination will be necessary.

continued

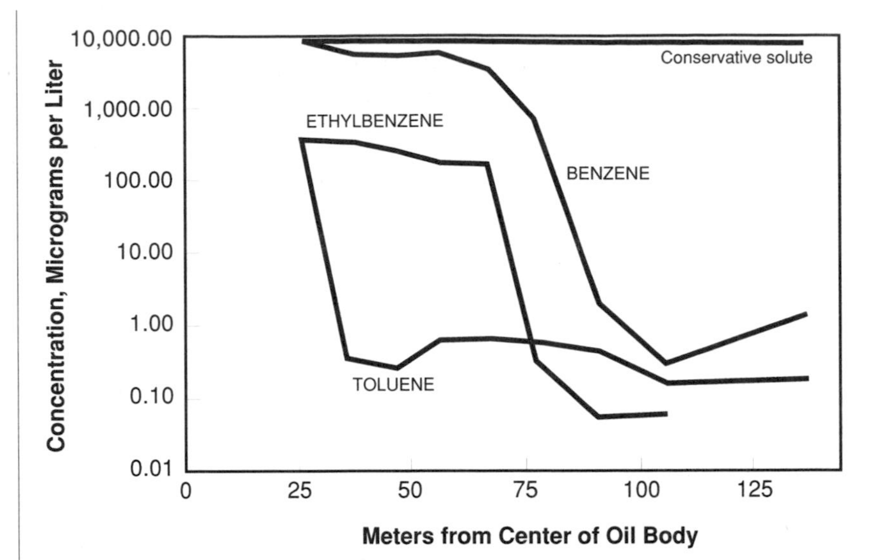

FIGURE 2-3 Concentrations of the crude oil components toluene, ethylbenzene, and benzene at various distances from the center of the Bemidji, Minnesota, oil spill. These concentrations have remained relatively stable at the levels shown here since 1987. Note that the contaminant concentrations are very high near the center of the plume but that they drop dramatically within 100 m of the spill. If the contaminants were not biodegradable, this concentration drop would not occur, and the contamination would have spread much farther, as shown by the hypothetical concentration of a nondegradable solute (called a "conservative solute") pictured here. Thus, at this site, intrinsic bioremediation has effectively confined the contamination to a small region. SOURCE: Baedecker et al. (in press), reprinted with permission.

References

Baedecker, M. J., D. I. Siegal, P. E. Bennett, and I. M. Cozzarelli. 1989. The Fate and Effects of Crude Oil in a Shallow Aquifer: I. The Distribution of Chemical Species and Geochemical Facies. U.S. Geological Survey Water-Resources Investigations Report 88-4220. Reston, Va.: U.S. Geological Survey.

Baedecker, M. J., I. M. Cozzarelli, D. I. Siegal, P. E. Bennett, and R. P. Eganhouse. In press. Crude oil in a shallow sand and gravel aquifer: 3. Biogeochemical reactions and mass balance modeling in anoxic groundwater. Applied Geochemistry.

Cozzarelli, I. M., R. P. Eganhouse, and M. J. Baedecker. 1991. Transformation of monoaromatic hydrocarbons to organic acids in anoxic groundwater environment. Environmental Geology and Water Sciences 16(2):135-141.

Hult, M. F. 1984. Ground-Water Contamination by Crude Oil at the Bemidji, Minnesota Research Site. U.S. Geological Survey Water-Resources Investigations Report 84-4188. Reston, Va.: U.S. Geological Survey.

have also been used to describe intrinsic bioremediation. Box 2-2 describes a Minnesota site where researchers documented that intrinsic bioremediation prevented the further spreading of crude oil contamination.

Engineered bioremediation is the acceleration of microbial activities using engineered site-modification procedures, such as installation of wells to circulate fluids and nutrients to stimulate microbial growth. The principal strategy of engineered bioremediation is to isolate and control contaminated field sites so that they become in situ bioreactors. Other terms used to describe engineered bioremediation include "biorestoration" and "enhanced bioremediation."

As summarized in Box 2-3 and described below, the site conditions that influence a bioremediation project's success differ for intrinsic and engineered bioremediation.

Site Conditions for Engineered Bioremediation

Because engineered bioremediation uses technology to manipulate environmental conditions, the natural conditions are less important for engineered than for intrinsic bioremediation. For engineered bioremediation, the critical property influencing success is how well the subsurface materials at the site transmit fluids. For systems that circulate ground water, the hydraulic conductivity (the amount of ground water that moves through a unit section of the subsurface in a given time) in the area containing the contaminant should be on the order of 10^{-4} cm/s or greater (the precise value is site specific). For systems that circulate air, the intrinsic permeability (a measure of how easily fluids flow through the subsurface) should be greater than 10^{-9} cm^2. For both types of systems, the contaminated area will be much more difficult to treat if it has crevices, fractures, or other irregularities that allow channeling of fluids around contaminated material. Land near river deltas, floodplains of large rivers, and areas where sand and gravel aquifers were formed from the melting of glaciers can be uniform over large areas. On the other hand, braided stream channels can contain a substantial number of irregularities that complicate bioremediation system design.

At high concentrations, contaminants (including petroleum products and chlorinated solvents) that form a nonaqueous-phase liquid can exclude water or air from pores in the subsurface. Nonaqueous-phase liquids restrict access of the remedial fluids and gases and complicate engineered bioremediation. In most cases such contaminants at residual concentrations of less than 8000 to 10,000 mg/kg of soil do not significantly affect water or air flow, because at this level

BOX 2-3
SITE CHARACTERISTICS THAT FAVOR IN SITU BIOREMEDIATION

Engineered bioremediation requires installing wells and other engineering systems to circulate electron acceptors and nutrients that stimulate microbial growth. Key site characteristics for engineered bioremediation are:

❑ Transmissivity of the subsurface to fluids:

- hydraulic conductivity greater than 10^{-4} cm/s (if the system circulates water)
- intrinsic permeability greater than 10^{-9} cm^2 (if the system circulates air)

❑ Relatively uniform subsurface medium (common in river delta deposits, floodplains of large rivers, and glacial outwash aquifers)

❑ Residual concentration of nonaqueous-phase contaminants of less than 10,000 mg/kg of subsurface solids.

Intrinsic bioremediation destroys contaminants without human intervention, as the population of native microbes capable of degrading the contaminant increases naturally. The process requires thorough site monitoring to demonstrate that contaminant removal is occurring. Key characteristics of sites amenable to intrinsic bioremediation are:

❑ Consistent ground water flow (speed and direction) throughout the seasons:

- seasonal variation in depth to water table less than 1 m
- seasonal variation in regional flow trajectory less than 25 degrees

❑ Presence of carbonate minerals (limestone, dolomite, shell material) to buffer pH

❑ High concentrations of electron acceptors such as oxygen, nitrate, sulfate, or ferric iron

❑ Presence of elemental nutrients (especially nitrogen and phosphorus)

the contaminants are essentially nonmobile and occupy much less pore space than the water. The specific concentration value at which nonaqueous-phase contaminants begin interfering with fluid circulation varies depending on the contaminant (the value is higher for denser contaminants) and the soil.

Site Conditions for Intrinsic Bioremediation

If intrinsic bioremediation is the only option, ambient site conditions must be accepted as constraints for meeting cleanup goals, because intrinsic bioremediation by definition occurs without adding anything to the site. Only a fraction of the contaminated sites offer naturally occurring hydrogeochemical conditions in which microorganisms can degrade contaminants quickly enough to prevent them from spreading without human intervention.

The critical site characteristic for intrinsic bioremediation is predictability of ground water flow in time and space. Predictable water flow is essential for determining whether the native microbes will be able to act in all the places where the contaminant might travel in all seasons and for determining whether the microbes can act quickly enough to prevent the contamination from spreading with the flowing ground water. The hydraulic gradient and trajectory of ground water flow should be consistent through the seasons and from year to year. To ensure predictability of flow, the fluctuation in the water table should not vary more than about 1 m, although the precise number is site specific. In addition, the trajectory of regional flow should not change by more than about 25 degrees from the primary flow direction. These circumstances are more likely in upland landscapes with humid, temperate climates. In contrast, contaminant plumes in estuaries or the flood plains of large rivers often behave unpredictably.

Another valuable characteristic is the presence in the aquifer of minerals such as carbonates to buffer pH changes that would otherwise result from biological production of carbon dioxide or other acids or bases. Carbonates in the aquifer matrix can be expected when limestone or dolomite are the parent material or when limestone dust or sand occurs in glacial outwash. Carbonates can also occur as shell material in beach deposits.

Intrinsic bioremediation is more extensive when the ambient ground water surrounding the spill has high concentrations of oxygen or other electron acceptors. The importance of ambient concentrations of nitrate, sulfate, and ferric iron as potential electron acceptors that can stimulate microbial growth in the absence of oxygen is too often ignored. Most ground waters have more nitrate and sulfate than oxygen. This is particularly true in agricultural areas that have been overfertilized and in arid regions where gypsum is dissolved in the ground water.

The concentration of electron acceptors required to ensure bioremediation varies with the contaminant's chemical characteristics and the amount of contamination. More soluble contaminants and large contaminant sources require larger electron acceptor concentrations. Natural ground water circulation conditions at the site also influence the required amount of electron acceptor. The circulation patterns should provide enough mixing between contaminated water and surrounding water that the organisms never consume all of the electron acceptors within the bioremediation region. If the electron acceptor supply becomes depleted, bioremediation will slow or cease.

Also necessary for intrinsic bioremediation is the presence of the elemental nutrients that microbes require for cell building, especially nitrogen and phosphorus. Although nutrients must be present naturally for intrinsic bioremediation to proceed, the quantity of nutrients required is much less than the quantity of electron acceptors. Therefore, a nutrient shortage is less likely to limit intrinsic bioremediation than an inadequate electron acceptor supply.

Impact of Site Heterogeneity on Bioremediation

Observation of the geological cross section at a typical excavation site reveals a complex patchwork of layers, lenses, and fingers of different materials. Indeed, two overriding characteristics of the subsurface are that it is intricately heterogeneous and difficult to observe. The patterns of variability of the properties that govern the flow of water and the transport of chemicals are so complex that it is not possible to predict these properties quantitatively or even to interpolate them with accuracy from sparse observations. In practice, estimation of subsurface hydrogeochemical properties depends on site-specific measurements from water or soil samples and well tests. However, the inherent unobservability of the system means that there is usually insufficient information to characterize the site with certainty.

A consequence of this complexity and heterogeneity, in combination with the poor observability of the subsurface, is that completely reliable prediction of chemical transport and fate is out of reach in most real-world cases. In evaluating a proposed intrinsic or engineered bioremediation scheme, one must consider how it may perform under variable and not perfectly known conditions. A scheme that works optimally under specific conditions but poorly otherwise may be inappropriate for in situ bioremediation.

FURTHER READING

While this chapter has briefly reviewed the principles underlying successful bioremediation, the references listed in Table 2-2 provide more thorough coverage of the key disciplines related to bioremediation. The list is not exhaustive. The references it provides were selected to represent the diversity of attitudes, perspectives, and paradigms that are pertinent to understanding bioremediation.

TABLE 2-2 Recommended Sources for Obtaining In-Depth Information About the Disciplines Pertinent to Bioremediation

Discipline	Reference	Synopsis
Environmental microbiology	Chapelle, F. H. 1993. Ground-Water Microbiology and Geochemistry. New York: John Wiley & Sons.	Reviews how the growth, metabolism, and ecology of microorganisms affect ground water chemistry in both pristine and chemically contaminated aquifer systems.
	Gibson, D. T. 1984. Microbial Degradation of Organic Compounds. New York: Marcel Dekker.	Provides a detailed survey of how microorganisms metabolize organic compounds. Each chapter, written by a different expert, focuses on a different class of compounds.
	Madsen, E. L. 1991. Determining in situ biodegradation: facts and challenges. Environmental Science and Technology 25:1661-1673.	Reviews principles and limitations of environmental microbiology as they apply to determining in situ biodegradation. Proposes useful approaches, especially as applicable to academic research.
	Madsen, E. L., and W. C. Ghiorse. 1993. Ground water microbiology: subsurface ecosystems processes. Pp. 167-213 in Aquatic Microbiology: An Ecological Approach, T. Ford, ed. Cambridge, Mass.: Blackwell Scientific Publishers.	Reviews major concepts and methodological developments that determine our understanding of microorganisms and the processes they carry out in subsurface ecosystems.
	VanLoosdrecht, M. C. M., J. Lyklema, W. Norde, and A. J. B. Zehnder. 1990. Influences of interfaces on microbial activity. Microbiological Reviews 54:75-87.	Provides a critical and cross-disciplinary review of how surfaces affect microbial activity and substrate availability.

Hydrogeology	Domenico, P. A., and F. W. Schwartz. 1990. Physical and Chemical Hydrogeology. New York: John Wiley & Sons.	Reviews principles and practice of ground water hydrology, with emphasis on environmental applications.
	Freeze, R. A., and J. A. Cherry. 1979. Groundwater. Englewood Cliffs, N.J.: Prentice-Hall.	Provides a comprehensive presentation of the theory, principles, and practice of hydrogeology.
Environmental engineering	McCarty, P. L. 1988. Bioengineering issues related to in situ remediation of contaminated soils and groundwater. Pp. 143-162 in Environmental Biotechnology: Reducing Risks from Environmental Chemicals Through Biotechnology, G. S. Omenn, ed. New York: Plenum Press.	Discusses engineering issues relevant to in situ bioremediation.
	Rittmann, B. E., A. J. Valocchi, E. Seagren, C. Ray, and B. Wrenn. 1992. A Critical Review of In Situ Bioremediation. Chicago: Gas Research Institute.	Provides a comprehensive critical review of the microbiological, engineering, and institutional possibilities and restrictions for in situ bioremediation.
Statistics	ASCE Task Committee. 1990. Review of geostatistics in geohydrology, parts I and II. ASCE Journal of Hydraulic Engineering 116(5):612-658.	Discusses geostatistical techniques and how they can assist in the solution of estimation problems, including interpolation, averaging, and network design.

continued

TABLE 2-2 *(continued)*

Discipline	Reference	Synopsis
Contaminant fate and transport	Fetter C. W. 1993. Contaminant Hydrogeology. New York: Macmillan Publishing Co.	Gives a comprehensive treatment of ground water contaminants and their transport, retardation, and transformation in the subsurface. Particularly strong on the chemistry of organic contaminants.
	National Research Council. 1990. Ground Water Models. Washington, D.C.: National Academy Press.	Provides a thorough review of the theory, use, limitations, and applications of computer modeling applied to the subsurface.
	Sahwney, B. L., and K. Brown, eds. 1989. Reactions and Movement of Organic Chemicals in Soils. Madison, Wisc.: Soil Science Society of America.	Provides a thorough review of sorption-desorption behavior of contaminants in soil. Includes chapters on contaminant movement and transformation.
Commercial application	Hinchee, R. E., and R. F. Olfenbuttel, eds. 1992. In Situ Bioreclamation: Applications and Investigations for Hydrocarbon and Contaminated Site Remediation. Boston: Butterworth-Heinemann.	Papers in this compendium discuss field and research studies of in situ and on-site bioremediation.

3

The Current Practice of Bioremediation

Increasingly, in situ bioremediation is being heralded as a promising "new" alternative ground water cleanup technology. In fact, however, bioremediation is not new. It has been used commercially for more than 20 years. The first commercial in situ bioremediation system was installed in 1972 to clean up a Sun Oil pipeline spill in Ambler, Pennsylvania.

Since 1972, bioremediation has become well developed as a means of cleaning up spills of gasoline, diesel, and other easily degraded petroleum products. In general, in situ bioremediation has not developed to the point where it can be used on a commercial scale to treat compounds other than easily degraded petroleum products. However, although in situ bioremediation of petroleum-based fuels is the only common use of the technology now, in the future bioremediation will likely be used to treat a broad range of contaminants. Recently, research has intensified on bioremediation of less easily degraded compounds, such as chlorinated solvents, pesticides, and polychlorinated biphenyls (PCBs). Bioremediation of many such recalcitrant compounds has been successfully field tested (see the case examples in Boxes 4-1 and 4-3, in Chapter 4).

This chapter describes the state of the practice of in situ bioremediation as used today. Although the current uses of bioremediation apply primarily to petroleum-based fuels, the principles of practice outlined here extend to a much broader range of uses for the technology in the future.

BIOREMEDIATION VERSUS OTHER TECHNOLOGIES

For the past decade, the method of choice for ground water cleanup has been pump-and-treat systems. These systems consist of a series of wells used to pump water to the surface and a surface treatment facility used to clean the extracted water. This method can control contaminant migration, and, if recovery wells are located in the heart of the contaminant plume, it can remove contaminant mass. However, recent studies have shown that because many common contaminants become trapped in the subsurface, completely flushing them out may require the pumping of extremely large volumes of water over very long time periods. In situ bioremediation, because it treats contaminants in place instead of requiring their extraction, may speed the cleanup process. Consequently, bioremediation is likely to take a few to several years compared to a few to several decades for pump-and-treat technology. Thus, while capital and annual operating costs may be higher for bioremediation, its shorter operating time usually results in a reduction of total costs. Factors contributing to cost reductions in bioremediation compared to pump-and-treat systems include reduced time required for site monitoring, reporting, and management, as well as reduced need for maintenance, labor, and supplies. Furthermore, the surface treatment methods that are part of pump-and-treat systems typically use air stripping and/or carbon treatment to remove contaminants from the water—processes that transfer the contaminant to another medium (the air or the land) instead of destroying it. Bioremediation, on the other hand, can completely destroy contaminants, converting them to carbon dioxide, water, and new cell mass.

For cleaning up contaminated soils, in situ bioremediation is only one of several possible technologies. Alternatives include (1) excavation followed by safe disposal or incineration, (2) on-site bioremediation using land-farming or fully enclosed soil cell techniques, (3) low-temperature desorption, (4) in situ vapor recovery, and (5) containment using slurry walls and caps. In situ methods (desorption, vapor recovery, containment, and bioremediation) have the advantages of being minimally disruptive to the site and potentially less expensive. Because ex situ methods require excavation, they disrupt the landscape, expose contaminants, and require replacement of soils. For these reasons, ex situ methods sometimes are impractical. Potential advantages of bioremediation compared to other in situ methods include destruction rather than transfer of the contaminant to another medium, minimal exposure of on-site workers to the contaminant, long-term protection of public health, and possible reduction in the duration of the remediation process.

BASICS OF BIOREMEDIATION PROCESS DESIGN

Biological and nonbiological measures to remedy environmental pollution are used the same way. All remediation techniques seek first to prevent contaminants from spreading. In the subsurface, contaminants spread primarily as a result of partitioning into ground water. As the ground water advances, soluble components from a concentrated contaminant pool dissolve, moving forward with the ground water to form a contaminant plume. Because the plume is mobile, it could be a financial, health, or legal liability if allowed to migrate off site. The concentrated source of contamination, on the other hand, often has settled into a fixed position and in this regard is stable. However, until the source can be removed (by whatever cleanup technology), the plume will always threaten to advance off site.

Depending on the nature of the site, the types of contaminants, and the needs of the parties responsible for the contaminated site, the treatment technologies administered may vary. The source area and the ground water plume may be treated by engineered bioremediation, intrinsic bioremediation, a combination of the two, or a mixture of bioremediation with nonbiological treatment strategies. Contaminant concentrations in ground water plumes are typically much lower than in the source area. Because of this concentration difference, management procedures for the source area and the plume may be quite different. When the source area is highly contaminated, aggressive containment and treatment are often required to bring the site under control.

Selection and application of a bioremediation process for the source or the plume require the consideration of several factors. The first factor is the goals for managing the site, which may vary from simple containment to meeting specific regulatory standards for contaminant concentrations in the ground water and soil. The second factor is the extent of contamination. Understanding the types of contaminants, their concentrations, and their locations is critical in designing in situ bioremediation procedures. The third factor is the types of biological processes that are effective for transforming the contaminant. By matching established metabolic capabilities with the contaminants found, a strategy for encouraging growth of the proper organisms can be developed. The final consideration is the site's transport dynamics, which control contaminant spreading and influence the selection of appropriate methods for stimulating microbial growth.

Once site characteristics have been discerned, strategies for gaining hydrologic control and for supplying the requisite nutrients and electron acceptors for the microorganisms can be developed. If there is a sufficient natural supply of these substances, intrinsic bioremediation may be effective. On the other hand, if these biochemical or environmental requirements must be artificially supplied to maintain a desired level of activity, engineered bioremediation is the desired course. The ultimate consideration is if and when the targeted cleanup goal can be achieved.

Engineered Bioremediation

Engineered bioremediation may be chosen over intrinsic bioremediation because of time and liability. Because engineered bioremediation accelerates biodegradation reaction rates, this technology is appropriate for situations where time constraints for contaminant elimination are short or where transport processes are causing the contaminant plume to advance rapidly. The need for rapid pollutant removal may be driven by an impending property transfer or by the impact of contamination on the local community. A shortened cleanup time means a correspondingly lower cost of maintaining the site, as more rapid remediation reduces the long-term sampling and analysis costs. Actions implicit in engineered bioremediation also address the political and psychological needs of a client or community that has been affected by the contamination. The construction and operation of engineered bioremediation systems can demonstrate to the local community that the party responsible for the contamination has a responsive "good neighbor" attitude.

Engineered bioremediation can take a number of forms. The different applications vary according to both the context of the contamination (site geology, hydrology, and chemistry) and the biochemical processes to be harnessed. Regardless of site conditions, however, certain principal parameters guide the design, depending on whether the system is to treat soil or water.

Bioremediation Systems for Unsaturated Soils

When subsurface contamination exists substantially or entirely above the water table (in what is known as the unsaturated, or va-

This test plot in Kalkaska County, Michigan, is being used to demonstrate a simple form of engineered bioremediation for soil. The soil is contaminated with crude oil from an oil well. Tilling provides oxygen, and periodic irrigation maintains the soil moisture. The black tarp shown here helps warm the soil, which speeds microbial activity and prevents rain from infiltrating the site and contaminating deeper levels of the subsurface. Cleanup began in September 1991. Within a year, total petroleum hydrocarbons dropped from tens of thousands of parts per million to less than 400 ppm. Researchers expect that the concentrations will drop further, to nondetectable levels, within one more growing season. Based on the success at this site, the state of Michigan plans to approve bioremediation as a cleanup method for soils contaminated with crude oil. CREDIT: John M. Shauver, Michigan Department of Natural Resources.

dose, zone), the treatment system relies on transport of materials through the gas phase. Thus, engineered bioremediation is effected primarily through the use of an aeration system, oxygen being the electron acceptor of choice for the systems used so far to treat petroleum contamination. If the contamination is shallow, simple tilling of the soil may accelerate oxygen delivery sufficiently to promote bioremediation. For deeper contamination, aeration is most commonly provided by applying a vacuum, but it may also be supplied by injecting air. In either case the three primary control parameters are, in order of importance, oxygen supply, moisture maintenance, and the supply of nutrients and other reactants. Figure 3-1 shows an

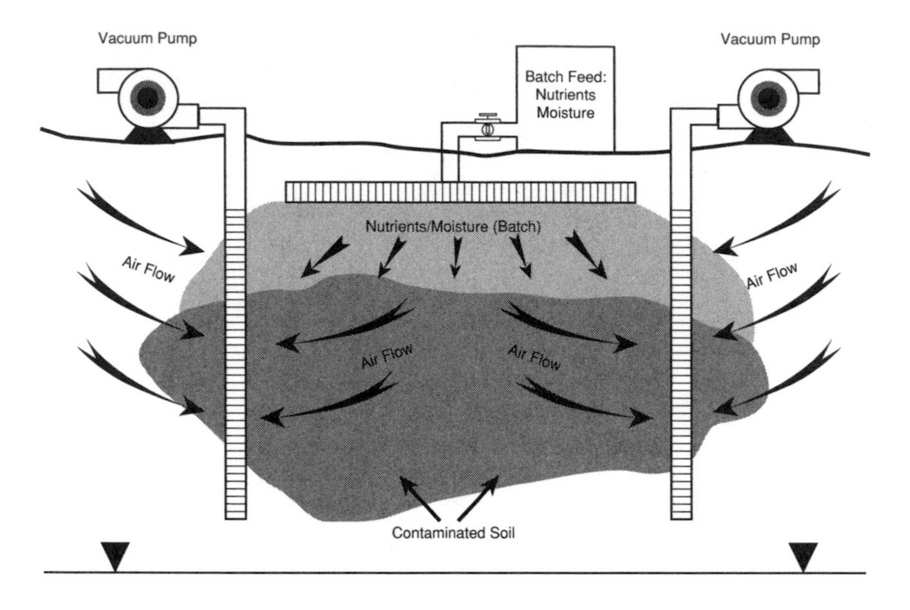

FIGURE 3-1 Engineered bioremediation system for treating soil above the water table (indicated by triangles at the bottom of the drawing). The vacuum pumps circulate air to supply oxygen. The infiltration gallery in the center of the diagram supplies water to replace lost moisture and nutrients to stimulate microbial growth.

engineered bioremediation system for unsaturated soils. It indicates the types of systems used to supply oxygen and nutrients and to maintain moisture.

The design and implementation of an effective vacuum or injection system for oxygen delivery require knowledge of the vertical and horizontal location of the contaminants and the geological characteristics of the contaminated zones. Because air flow is proportional to the permeability characteristics of each geological stratum, aeration points must be separately installed at depths that correspond to every contaminated geological unit. For effective oxygen delivery, the spacing of the aeration points within a geological unit is a function of the soil permeability and the applied vacuum (or pressure). Determination of spacing should be based on field data and/or computer models. In some clay-rich soils the circulation of sufficient oxygen to promote bioremediation is extremely difficult because such soils are relatively impermeable. In these soils hydraulic fracturing or another engineered approach may be required to facilitate air flow.

The passage of air through the subsurface will remove moisture.

This can cause drying that, if severe enough, may impede biological processes. Therefore, maintaining a proper moisture balance is critical to the system's success. Moisture is added to the treatment area by spraying or flooding the surface (if the surface is relatively permeable) or by injecting water through infiltration galleries, trenches, or dry wells. Care must be taken that excess water is not added, because it can leach contaminants into the ground water or decrease the amount of air in the subsurface pores.

If inorganic nutrients or other stimulants are required to maintain the effectiveness of the bioremediation system, they may be added in soluble form through the system used for moisture maintenance, as shown in Figure 3-1. In some cases, nutrients and stimulants could be added as gases. At some sites, nitrogen has been added in the form of gaseous ammonia. Future applications of bioremediation could add methane gas to stimulate the cometabolism of chlorinated ethenes. Gaseous additives can be administered through wells or trenches constructed parallel to the aeration system.

Bioremediation Systems for Ground Water

Bioremediation systems for treating ground water below the water table fit two categories: water circulation systems and air injection systems. Most aquifer bioremediation systems have used the former approach, but in the last few years air injection systems have become increasingly common.

Water Circulation Systems. Water circulation systems work by circulating water amended with nutrients and other substances required to stimulate microbial growth between injection and recovery wells. This approach is sometimes referred to as the Raymond Method, named after the scientist who designed the system for the 1972 Sun Oil spill. The method has typically incorporated an optional above-ground water treatment facility into the ground water circulation system. Figure 3-2 shows a diagram of a water circulation system, with oxygen supplied by hydrogen peroxide (H_2O_2) and the recovered water treated with an air stripper to remove any remaining volatile contaminants.

Under normal operations, all of the ground water is recovered, and all or a portion of the treated ground water is reinjected after being amended with nutrients and a final electron acceptor. Recovery systems most frequently use wells, although trenches can be used in some situations. Injection is commonly achieved with wells, but several systems have used injection galleries. In some systems all of the recovered water is discharged to an alternate reservoir, and ei-

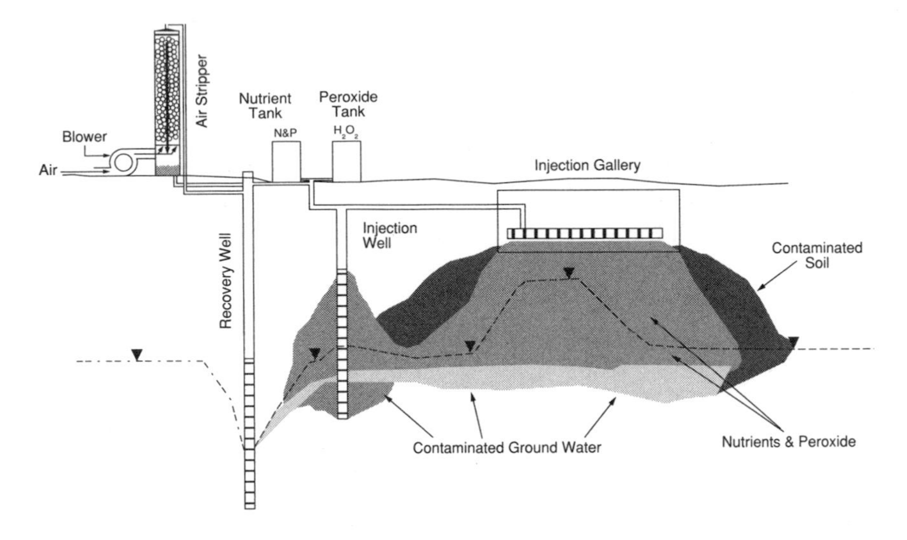

FIGURE 3-2 Water circulation bioremediation system for treating contaminated ground water. Water containing H_2O_2 (as an oxygen source) and nutrients is circulated through the site to stimulate microbial growth. An air stripper treats the recovered water to remove remaining volatile contaminants.

ther drinking water or uncontaminated ground water is used for injection. The injected ground water moves through the saturated sediments toward the ground water capture system. As the amended water moves through the contaminated portions of the site, it increases microbial activity by providing the elements that limit intrinsic biodegradation.

At the Wexford County, Michigan, gas-processing plant pictured on the right, a water circulation bioremediation system is being used to clean up ground water contaminated with gasoline spilled from a tanker truck. The top photo shows the system that supplies oxygen to the microbes. Pure oxygen from the oxygen tank is bubbled into water moving through the pipes shown here to a series of injection wells. Originally, the water was also amended with nutrients, but tests showed no increase in the biodegradation rate with nutrient addition, so nutrient addition was stopped. The cylinder in the center of the photo removes iron from the water to prevent it from clogging the injection wells. The lower photo shows some of the monitoring wells (the capped cylinders) used to test the system's effectiveness. Treatment began in August 1991. Within 14 months, pollutant concentrations dropped to levels that met state standards. CREDIT: John M. Shauver, Michigan Department of Natural Resources.

NO SMOKING

Nutrients typically added are nitrogen and phosphorus, although other minerals are occasionally used. Ammonium and nitrate salts are the most common nitrogen sources. Orthophosphate and tripolyphosphate salts are the most common phosphorous sources. The electron acceptor is most commonly oxygen in the form of air, pure oxygen, or hydrogen peroxide. Nitrate has also been used as an electron acceptor in some commercial-scale engineered bioremediation systems.

Air Injection Systems. Historically, one of the major limitations of water circulation systems has been the effective supply of an electron acceptor. Delivery of oxygen, the most common electron acceptor, is difficult because oxygen gas has limited water solubility and other oxygen vehicles (such as hydrogen peroxide and liquid oxygen) are costly and have had limited effectiveness. The difficulties associated with oxygen delivery have hampered the performance of bioremediation technology.

In the past few years, U.S. contractors have adopted the European practice of air sparging—the injection of air directly into ground water (see Figure 3-3). Air sparging serves two purposes. First, it is an efficient method of delivering oxygen to promote microbial growth. The injected air displaces water in the subsurface, creating pores tem-

At the Hanahan, South Carolina, petroleum tank farm pictured on the left, a water circulation bioremediation system is being used to clean up extensive ground water contamination from leaks in storage tanks and disposal of tank bottoms. The site contains a mixture of a wide variety of petroleum hydrocarbons, including aliphatic hydrocarbons, polycyclic aromatic hydrocarbons (PAHs), and gasoline components (benzene, ethylbenzene, toluene, and xylene, or BTEX). The treatment system consists of infiltration galleries used to circulate water amended with nitrate, which serves as an electron acceptor.

The top photo shows the tank farm. The lower photo shows construction of an infiltration gallery. The perforated pipe in the gravel-lined trench will be used to deliver the nutrients and water. The trench will be packed with gravel and capped with sand after the pipe is installed.

The treatment goals at this site are to destroy PAHs and aliphatic hydrocarbons sorbed to the soil and to decrease the BTEX concentration in the water. Researchers will use a subsite in which water has been infiltrated but no nitrate has been added as a control. The control subsite will be used to determine the effect of nitrate addition, and the resultant stimulation of microorganisms, on the rate of contaminant destruction.

CREDIT: Don A. Vroblesky, U.S. Geological Survey.

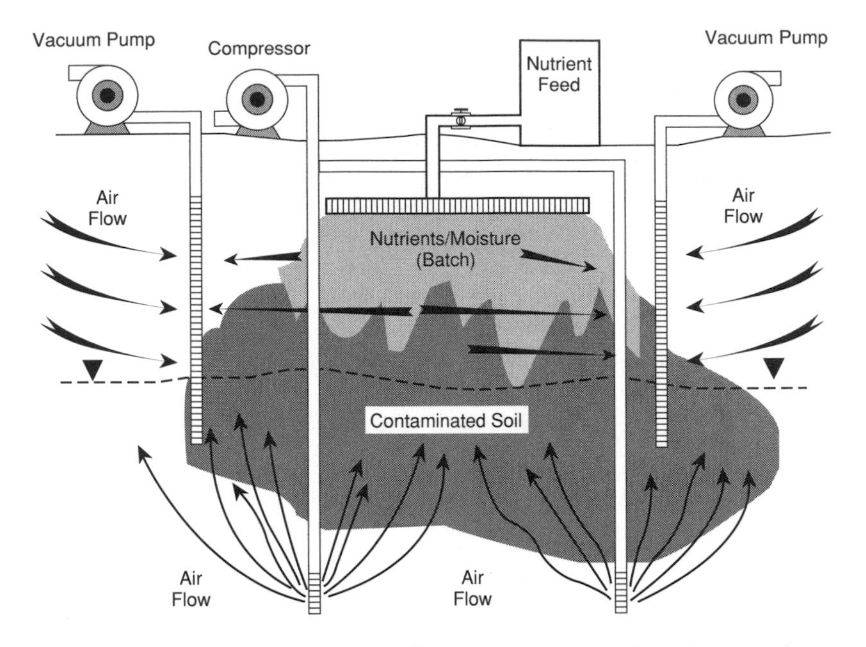

FIGURE 3-3 Air-sparging system for treating contaminated ground water. Vacuum pumps circulate air to promote the growth of aerobic microbes and to extract volatile contaminants. An infiltration system supplies nutrients.

porarily filled with air that is easily available to the microbes. Second, air sparging can help remove volatile contaminants. As the injected air sweeps upward through the contaminated zone, it can carry volatile contaminants to the soil above the water table for capture by a vapor recovery system.

Essential to the proper use of air sparging is delineation of the extent of contamination and the subsurface geological profile. These requirements must be met to ensure that air can move readily and uniformly through the area to be treated. If there are geological barriers that can trap or channel the air flow, the use of air sparging may be precluded. The spacing of wells and the injection pressure for the sparging system need to be determined by field testing and/or modeling.

Addition of nutrients or other amendments, if necessary, can be accomplished through the use of injection wells or infiltration galleries (as shown in Figure 3-3). The movement of air through the subsurface provides a mixing function that helps disperse nutrients through the water column. Gaseous reactants, such as methane, that may be required for cometabolic bioremediation strategies could be added through the sparge wells.

In most situations, sparging is conducted with a ground water capture system to prevent migration of dissolved contaminants. When volatile compounds are present, air recovery systems are used to prevent contamination of the air above the system and contaminant transfer to adjacent areas.

Intrinsic Bioremediation

Because intrinsic bioremediation relies on the innate capabilities of naturally occurring microbial communities, the capacity of the native microbes to carry out intrinsic bioremediation must be documented in laboratory tests performed on site-specific samples. These tests must be carried out before intrinsic bioremediation can be proposed as a legitimate cleanup strategy. In addition, the effectiveness of intrinsic bioremediation must be proven with a site-monitoring regime that includes chemical analysis of contaminants, final electron acceptors, and/or other reactants or products indicative of biodegradation processes (as explained in Chapter 4).

Intrinsic bioremediation may be used alone or in conjunction with other remediation techniques. For instance, soils may be excavated for disposal or treatment, with intrinsic bioremediation used to eliminate residual contamination. Similarly, this process may be implemented after a pump-and-treat or engineered bioremediation system has reduced the potential for migration of contaminants off site.

For intrinsic bioremediation to be effective, the rate of contaminant biodegradation must be faster than the rate of contaminant migration. These relative rates depend on the type of contaminant, the microbial community, and the subsurface hydrogeochemical conditions. Frequently, the rate-controlling step is the influx of oxygen. Lack of a sufficiently large microbial population can also limit the cleanup rate; this can be caused by a lack of nutrients or an inhibitory condition, such as low pH or the presence of a toxic material.

Prior to implementation of intrinsic bioremediation, the site must be thoroughly investigated. Parameters of concern include the type, mass, and distribution of contaminant; the contaminant's susceptibility to biodegradation by microorganisms at the site; the flow of ground water under nonpumping conditions (including seasonal fluctuations); historical data on plume migration; and the closeness and sensitivity of potential receptors that may be adversely affected if reached by the contaminant. If information on all of these parameters is available, a mathematical model can be used to predict the rates of migration and biodegradation. Thus, prospects for expansion or recession of the contaminated area can be assessed.

For regulators to approve intrinsic bioremediation, the regulatory agency must be provided with supportive data to ensure that the public health will be adequately protected. The implementation plan must include a site-monitoring program to confirm that intrinsic bioremediation is performing as expected. If performance falls short of expectations and the contamination spreads, further corrective action will likely be required.

INTEGRATION OF BIOREMEDIATION WITH OTHER TECHNOLOGIES

As a contaminated site is cleaned up, the biological reactions promoted in bioremediation affect site chemistry in ways that may make the site more amenable to cleanup with nonbiological technologies. For example, biological activity may speed contaminant desorption from solids, making it easier to extract the contaminants with a pump-and-treat system. Similarly, nonbiological cleanup technologies can affect microbial activity at a site, sometimes promoting bioremediation. For example, techniques designed to vent volatile contaminants may increase the oxygen supply, encouraging microbial growth. Such synergistic effects can maximize rates of contaminant loss. Thus, bioremediation is frequently integrated with other technologies, both sequentially and simultaneously. Some examples follow:

- When contaminant concentrations are high and affected zones are accessible, bioremediation frequently follows excavation of soils near the contaminant source. Excavated soils may be disposed of off site or treated by a surface (ex situ) bioremediation system or by a thermal method. Removal of heavily contaminated soils reduces the demands on an in situ technology and immediately reduces the potential for impact on the ground water.
- Where residual pools of contaminants are present in the subsurface, these pools may be removed prior to implementation of other remediation technologies by a process known as free product recovery. For pools of contaminants that are less dense than water, such as gasoline and diesel fuel, free product recovery is accomplished by pumping the liquids from wells or trenches. This removes the contaminant mass in the most concentrated form and reduces the demand for nutrients and electron acceptors during bioremediation procedures that may follow. No good recovery methods exist for pools of contaminants that are more dense than water, such as chlorinated solvents, because these tend to sink deep into the subsurface, where they are difficult to locate.

- Bioremediation may follow the use of a pump-and-treat system. The pump-and-treat system may be used to shrink the contaminant plume and, along with a free product recovery system, help remove contaminant mass. Once a sufficient quantity of the contaminant has been removed, the pump-and-treat system may be augmented with or converted to an in situ bioremediation system.
- Frequently, in situ bioremediation for cleaning up ground water is conducted in conjunction with in situ vapor recovery for cleaning up regions above the water table. The in situ vapor recovery system uses a series of recovery wells or trenches to extract air and volatile contaminants from above the water table. In addition to withdrawing volatile contaminants, the wells and trenches provide oxygen for biodegradation. Hence, the process has become commonly known as bioventing. By a combination of volatilization and biodegradation, contaminant levels above the water table are reduced, thus decreasing the potential for the contaminants to leach into the ground water. Further, as the water table drops during dry seasons, more subsurface sediments are exposed to air movement; thus, contamination is reduced within the zone of water table fluctuation. Alternatively, air sparging may be used along with vapor extraction procedures to reduce contamination below the water table.
- It is possible to follow engineered bioremediation with intrinsic bioremediation. After removal of free product contaminants, engineered bioremediation may be used to eliminate the majority of residual contaminants. Then, after an asymptotic decline of contaminants in the plume, final polishing and containment may be accomplished using intrinsic bioremediation. In this case, microbial activity will have been stimulated and the biodegradation process at the site will be well understood. (See Box 4-2, in Chapter 4, for an example.)

GOOD PRACTICES

The general approach required to earn credibility in the bioremediation industry is the same as for any technical business: work only within areas of expertise, be aware of the general limitations of the technology, pay attention to details, and work closely with clients. In general, buyers of bioremediation services can determine whether a bioremediation contractor is competent to do the job by reviewing the contractor's credentials. Competent contracting firms have employees and consulting experts with credentials in the scientific and engineering fields important to bioremediation. And, like any other successful business, a bioremediation firm should have a track record of reliable performance that can be determined by reviewing the firm's references. Box 3-1 lists standards of practice that all contractors should follow.

BOX 3-1
STANDARDS OF PRACTICE FOR BIOREMEDIATION CONTRACTORS

❑ Contractors should employ experts in the many scientific and engineering fields important in bioremediation, including environmental engineering, hydrogeology, microbiology, and chemistry.

❑ Contractors should be frank and open with the client about all uncertainties in the process.

❑ Before treatment begins, the contractor should negotiate with all involved parties (clients, regulators, and the affected community) the standards that will be used to evaluate process performance. Agreeing on performance standards will prevent conflicts resulting from transient or trace amounts of contaminants found at sites after the treatment is completed and will give the clients a realistic picture of what to expect after the project is finished.

❑ The contractor should develop a clearly defined engineering design for the treatment program. The design should provide a clear course toward achieving a specific endpoint.

❑ The contractor's design should be based on site-specific data.

❑ The design should include a clearly defined monitoring program.

❑ The design should leave room for flexibility based on operational data that indicate a need for adjustments, especially if the process is innovative. Operators should be informed of the need to adjust the system.

4

Evaluating In Situ Bioremediation

Showing that a bioremediation project is working requires evidence not only that contaminant concentrations have decreased but also that microbes caused the decrease. Although other processes may contribute to site cleanup during a bioremediation, the microbes should be critical in meeting cleanup goals. Without evidence of microbial involvement, there is no way to verify that the contaminant did not simply volatilize, migrate off site, sorb to subsurface solids, or change form via abiotic chemical reactions. This chapter discusses a strategy for evaluating the effectiveness of in situ bioremediation projects, based on showing that microbes were responsible for declining contaminant concentrations. Regulators and buyers of bioremediation services can use the strategy to evaluate the soundness of a proposed or ongoing in situ bioremediation system. Researchers can apply the strategy to evaluate the results of field tests.

A THREE-PART STRATEGY FOR "PROVING" IN SITU BIOREMEDIATION

To answer the question "What proves in situ bioremediation?", one must recognize that only under rare circumstances is proof of in situ bioremediation unequivocal. In the majority of cases the complexities of contaminant mixtures, their hydrogeochemical settings, and competing abiotic mechanisms of contaminant loss make it a challenge to identify biodegradation processes. Unlike controlled laboratory experiments in which measurements can usually be interpreted easily, cause-and-effect relationships are often difficult to establish at field sites. Furthermore, certain data that may convince some authorities of in situ bioremediation may not convince others.

Although proving microbial involvement in cleanup with complete certainty is seldom possible, the weight of the evidence should point to microbes as key actors in the cleanup. Because one measurement is seldom adequate, the evaluation strategy must build a consistent, logical case that relies on convergent lines of independent evidence taken from the field site itself. The general strategy for demonstrating that in situ bioremediation is working should include three types of evidence:

1. documented loss of contaminants from the site,
2. laboratory assays showing that microorganisms in site samples have the *potential* to transform the contaminants under the expected site conditions, and
3. one or more pieces of evidence showing that the biodegradation potential is *actually realized* in the field.

The strategy applies not only to bioremediation projects in the implementation phase but also to those in the testing phase. The strategy is not just for research purposes. Every well-designed bioremediation project should show evidence of meeting the strategy's three requirements. Thus, regulators and buyers of bioremediation services can use the strategy to evaluate whether a proposed or ongoing bioremediation project is sound.

The first type of evidence in the strategy—showing decreasing contaminant concentrations—comes from standard sampling of the ground water and soil over time as cleanup progresses. The second type of evidence—showing the potential for microorganisms to degrade the contaminants—is also relatively simple to provide. In most cases it requires taking microbes from the field and showing that they can degrade the contaminant when grown in a well-controlled laboratory vessel. For some cases, lab studies may not be needed when a body of peer-reviewed published studies documents that the compounds are easily and commonly biodegraded.

The most difficult evidence to gather is the third type—showing that the biodegradation potential demonstrated in the laboratory is being realized in the field. Evidence of field biodegradation is essential: data showing that organisms are capable of degrading the contaminant in the laboratory are not sufficient because the organisms may not perform the same tasks under the less hospitable field conditions. A variety of techniques, explained below, exist for demonstrating field biodegradation.

TECHNIQUES FOR DEMONSTRATING BIODEGRADATION IN THE FIELD

The goal of the techniques for demonstrating field biodegradation is to show that characteristics of the site's chemistry or microbial population change in ways that one would predict if bioremediation were occurring. The environmental changes measured in these tests should correlate to documented contaminant loss over time. No one technique alone can show with complete certainty that biodegradation is the primary reason for declining contaminant concentrations in the field. The wider the variety of techniques used, the stronger the case for successful bioremediation. As an example, Box 4-1 describes a site where several tests were combined.

There are two types of sample-based techniques for demonstrating field biodegradation: measurements of field samples and experiments run in the field. In most bioremediation scenarios a third technique, modeling experiments, provides an improved understanding of the fate of contaminants. Examples of field measurements, field experiments, and modeling experiments are described below. These examples provide general guidance about which types of tests are appropriate. Detailed experimental protocols for carrying out the tests need to be developed and will vary depending on the types of contaminants present, the geological characteristics of the site, and the level of rigor desired in the evaluation.

Measurements of Field Samples

A number of techniques for documenting in situ bioremediation involve removing samples of soil and water from the site and bringing them to the lab for chemical or microbiological analysis. Many of these techniques require comparing conditions at the site once bioremediation is under way with site conditions under baseline circumstances, when bioremediation is not occurring. Baseline conditions can be established in two ways. The first method is to analyze samples from a location that is hydrogeologically similar to the area being treated but is either uncontaminated or is outside the zone of influence of the bioremediation system. The second method is to gather samples before starting the bioremediation system and compare them with samples gathered at several time points after the system is operating. This second method applies only to engineered

BOX 4-1
PROVING ENGINEERED BIOREMEDIATION OF CHLORINATED SOLVENTS IN A FIELD TEST—MOFFETT NAVAL AIR STATION, CALIFORNIA

Researchers at Stanford University conducted a field demonstration to evaluate the potential for using cometabolism for in situ bioremediation of chlorinated solvents. The work done in this field demonstration shows how a variety of tests can be combined to evaluate whether new bioremediation processes researched in the lab can be applied successfully in the field.

The demonstration site was a highly instrumented and well-characterized confined sand-and-gravel aquifer at Moffett Naval Air Station in Mountain View, California. To this site the researchers purposely added chlorinated solvents in a carefully controlled way that ensured that the solvents would not migrate beyond the research plot. Chlorinated solvents cannot support microbial growth on their own, but, if supplied with methane, a special class of organisms can destroy the contaminants through cometabolism (see Chapter 2). Thus, at this site, researchers stimulated native organisms by adding oxygen and methane. When stimulated, the organisms destroyed significant quantities of the chlorinated solvents. The researchers evaluated the success of their work using tests that meet the three criteria discussed in this chapter:

1. *Documented loss of contaminants:* The researchers documented that 95 percent of the vinyl chloride, 85 percent of the *trans*-1,2-dichloroethylene, 40 percent of the *cis*-1,2-dichloroethylene, and 20 percent of the trichloroethylene added to the site were transformed.

2. *Laboratory assays showing that microorganisms have the potential to degrade the contaminants:* When cores of the aquifer removed to the lab were exposed to methane and oxygen, the methane and oxygen were used up, showing that the cores contained bacteria that thrive on methane (methanotrophs). Previous research had shown that methanotrophs can cometabolize chlorinated solvents.

3. *Evidence that biodegradation potential is realized in the field:* The researchers used a variety of methods to demonstrate biodegradation in the field. First, they showed that before the methanotrophs were stimulated with methane and oxygen, destruction of trichloroethylene was minimal. Second, they performed conservative tracer tests with bromine to show that the methane and oxygen added to the site were not disappearing by physical transport but were being used by microorganisms. Third, they identified microbial breakdown products from the solvents in aquifer samples. Fourth, they used models to show that theoretical estimates of biodegradation rates could account for contaminant loss in the field.

References

Roberts, P. V., G. C. Hopkins, D. M. Mackay, and L. Semprini. 1990. A field evaluation of in-situ biodegradation of chlorinated ethenes: Part 1—Methodology and field site characterization. Groundwater 28:591-604.

Semprini, L., P. V. Roberts, G. D. Hopkins, and P. L. McCarty. 1990. A field evaluation of in-situ biodegradation of chlorinated ethenes: Part 2—The results of biostimulation and biotransformation experiments. Groundwater 28:715-727.

bioremediation systems, because the "starting time" of intrinsic bioremediation occurs whenever the contaminant enters the subsurface and therefore cannot be controlled.

Following are several types of analyses that may be performed on samples removed from the field.

Number of Bacteria

When microbes metabolize contaminants, they usually reproduce. (In general, the larger the number of active microbes, the more quickly the contaminants will be degraded.) Thus, samples correlating contaminant loss with an increase in the number of contaminant-degrading bacteria above the normal conditions provide one indicator that active bioremediation may be occurring in the field. When contaminant biodegradation rates are low, such as when contaminant levels are low or biodegradable components are inaccessible, increases in the number of bacteria may not be great enough to detect above background levels, given the error in sampling and measurement techniques. Thus, the absence of a large increase in bacterial numbers does not necessarily mean that bioremediation is unsuccessful.

The first issue for determining the size of the bacterial population is what to sample. In principle, the best samples include the solid matrix (the soil and rock that hold the ground water) and the associated pore water. Because most microorganisms are attached to solid surfaces or are trapped in the intersticies between soil grains, sampling only the water normally underestimates the total number of bacteria, sometimes by as much as several orders of magnitude. In addition, water samples may misrepresent the distribution of microbial types, because a water sample may contain only those bacteria easily dislodged from surfaces or that can be transported in the moving ground water.

While obtaining soil samples from the earth's surface is not difficult, subsurface sampling is expensive and time consuming. Subsurface samples are obtained by removing cylindrical cores from below ground. A major effort is required to prevent microbial contamination of the sample during the coring operation and while handling the sample. Whenever possible, sampling equipment should be sterilized before use. Contamination from extraneous material, including the air, soil, and human contact, should be prevented.

Although ground water samples have important deficiencies, they have a role as semiquantitative indicators of microbial numbers. Major increases in the number of bacteria in the ground water usually correlate to large increases in the total number of bacteria in the subsurface. The main advantages of ground water samples are that they can be taken repeatedly from the same location and that they are relatively inexpensive.

The second issue for determining bacterial numbers is how to assay for the bacteria. Several standard and emerging techniques, each of which has advantages and disadvantages, are available:

- *Direct microscopic counting* is a traditional technique that involves using a microscope to view the sample and count the bacteria, which are distinguished from solid debris based on their size and shape. Microscopic counting is greatly aided by the use of the acridine orange stain and epifluorescence microscopy, which make intact bacteria stand out from other particles. Microscopic enumeration can be tedious and requires technician experience, particularly when the sample contains solids. The technique provides data on total bacterial counts but does not give information on cell types or metabolic activity.
- *The INT activity test* can enhance direct microscopic counting by identifying only those bacteria active in electron transport, the main force behind all metabolism. If the sample (or bacteria harvested from the sample) is incubated with a tetrazolium salt under controlled conditions, actively respiring bacteria transfer electrons to the tetrazolium salt, forming purple INT crystals that can be observed microscopically inside the metabolically active bacteria.
- *Plate counts,* another standard technique, can be used to quantify the number of bacteria able to grow on a prescribed set of nutrients and substrates immobilized in a solid medium. The solid medium is created from a liquid solution with the appropriate nutrients and substrates, solidified with agar to form a gel. A sample containing the bacteria of interest is spread thinly over the surface of the gel. After the plate is incubated, visible bacterial colonies form, and the colonies can be counted to indicate the number of metabolically ac-

tive bacteria in the original sample. Because plate counting requires significant growth to form visible colonies, the method often underestimates the number and diversity of bacteria. On the other hand, considerable information on the metabolic capability of bacteria can be obtained by using a range of growth substrates to prepare the media.

- *The most-probable-number (MPN) technique* also relies on significant growth in prescribed media, but enumeration is carried out through a statistical analysis. Instead of counting colonies for a few incubations on solid media, the MPN technique uses a large number of incubations from portions of the sample diluted to prescribed levels in a nutrient solution. Based on simple statistics and the number of diluted liquid samples that show evidence of bacterial growth, the number of bacteria in the original undiluted sample can be calculated. Although details of the MPN and plate-counting methods differ, both techniques have the same general advantages and disadvantages.

Modern tools of biochemistry and molecular biology are becoming available to provide more precise ways of identifying and enumerating bacteria in site samples. The tools exploit the growing understanding of genetically determined characteristics of particular cell components:

- *Oligonucleotide probes* are small pieces of deoxyribonucleic acid (DNA) that can identify bacteria by the unique sequence of molecules coded in their genes. The small DNA probe bonds with a complementary region of the target cell's genetic material, and the amount of bound probe can be quantified and correlated with the number of cells. Advanced probing techniques to count the cells in intact samples are under development. Probing is a very powerful technique for identifying which types of bacteria are present, as long as the genetic sequences are known for the target bacteria. This knowledge is available for some common types of bacteria and often is known for genetically engineered microorganisms or other specialized microorganisms that might be used as part of a bioaugmentation strategy. Probing also can be used to determine whether the gene for a particular biodegradation reaction is present. The drawbacks of probing are that it requires considerable prior knowledge of the cells' genetic sequences, it is only semiquantitative in its current state, and it requires specialized equipment and knowledge.
- *Fatty acid analysis* is a second new bacterial identification technique; it uses the characteristic "signature" of fatty acids present in the membranes of all cells. The distribution of fatty acids is unique

and stable among different bacteria and therefore can be used as an identifying signature. Like gene probing, fatty acid analysis requires specialized knowledge and equipment. It is limited in its quantification ability and may have sensitivity limitations for small populations.

With gene probing and fatty acid analysis, it is unnecessary to grow the bacteria in the laboratory to detect what kinds and how many are present in a sample. While holding great promise, these new methods are still in the development and testing stages.

Number of Protozoans

Because protozoans prey on bacteria, an increase in the number of protozoans suggests a major increase in the number of bacteria. Therefore, samples correlating contaminant loss with growth in the protozoan population can provide further evidence of active bioremediation. The protozoan population can be counted using a statistical MPN technique similar to that used for bacteria. The MPN technique for counting protozoans requires growing various dilutions of the soil or water sample in cultures containing a large number of bacterial prey. Whether protozoans grow to feed on these prey can be determined by viewing the diluted samples through a microscope.

Rates of Bacterial Activity

While an increase in bacterial numbers usually is a key sign that bioremediation is working, the stronger measure of success is that potential biotransformation rates are great enough to remove the contaminant rapidly or to prevent contaminant migration. Therefore, measurements demonstrating that the bacteria are capable of performing the desired reactions at significant rates help to provide evidence of successful bioremediation.

The most direct means for estimating biodegradation rates is to construct laboratory microcosms with environmental conditions as close as possible to those from which the sample was taken. (See Box 4-2 for an example of the use of microcosms.) To these microcosms, field samples and the accompanying microbes (or other microbes that could be released into the field) are added. Microcosms are useful because substrate concentrations and environmental conditions can be controlled and the loss of the contaminant or other markers of biodegradation can be measured relatively easily. For many pollutants (including BTEX and PCBs), versions labeled with carbon-14 are

BOX 4-2
PROVING ENGINEERED BIOREMEDIATION OF AN OIL AND FUEL SPILL—DENVER, COLORADO

In Denver, Colorado, a temporary holding tank under a garage used to service vehicles leaked crankcase oil, diesel fuel, and gasoline. The leak contaminated the surrounding soil and created a plume of benzene, toluene, ethylbenzene, and xylene (BTEX) in the ground water below. An engineered bioremediation system was installed at the site in 1989. The soil was cleaned by removing the remaining pools of leaked contaminants and by venting to supply oxygen and promote biodegradation. The ground water was treated by circulating oxygen (in the form of hydrogen peroxide), phosphorus (in the form of phosphate), and nitrogen (in the form of ammonium chloride) to promote bioremediation.

By March 1992, after three years of treatment, the dissolved plume of contaminants had been nearly eliminated from the ground water. However, tests revealed that the aquifer contained a small layer that had trapped considerable quantities of BTEX. This layer is relatively impermeable and therefore had been bypassed by the fluids circulated to promote bioremediation. When the bioremediation system was shut down in 1992, long-term monitoring began to ensure that the native microbial community could act quickly enough to degrade any contamination that might leak from this layer.

Although the engineered bioremediation system at this site was unable to eliminate all of the contamination, it succeeded in reducing the amount and risk of the contamination to acceptable levels. Furthermore, it is likely that microbes at the periphery of the remaining contamination will provide effective intrinsic bioremediation that will prevent the reemergence of a contaminant plume.

The cleanup using the engineered bioremediation met the three criteria set forth in this report:

1. *Documented loss of contaminants:* At the monitoring well closest to the gallery used to deliver oxygen and nutrients to the site, the BTEX concentration dropped from 2030 μg/l before bioremediation to 6 μg/l after bioremediation. At other monitoring wells, the concentration dropped more than an order of magnitude, to less than 46 μg/l.

2. *Laboratory assays showing that microorganisms have the potential to degrade the contaminants:* Studies showed that microorganisms in the transmissive layers adjacent to the trapped contaminants could consume as much as 7 mg/l of oxygen per day, resulting in the potential destruction of as much as 2 mg/l of hydrocarbons per

continued

day. This oxygen consumption rate was determined by placing a dewatered core from the site in a sealed glass mason jar and measuring the amount of oxygen the microbes in the core consumed in 24 hours. No direct tests—other than measuring the oxygen consumption rate—of the native microbes' ability to degrade BTEX were performed. However, the ability of subsurface microorganisms to degrade BTEX is well established (see Table 2-1), so direct lab tests were not as important for this site as for sites with contaminants for which bioremediation techniques are still emerging.

3. *Evidence that biodegradation potential is realized in the field:* At this site, two types of tests provided evidence of biodegradation in the field. First, the oxygen consumption rate in microcosms constructed with cores from the site was highest when the cores came from near the layer of trapped contaminants. Thus, microbes with access to the largest supply of contaminants consumed oxygen most rapidly, supporting the expectation that bacterial growth on the hydrocarbons had been stimulated. Second, the ratio of BTEX to total petroleum hydrocarbons (TPH) was lower in the bioremediated area than in the contaminant source. Research has shown that microorganisms prefer BTEX to other components of TPH, leaving a TPH residual that is relatively low in BTEX after a successful remediation.

Reference

Nelson, C., R. J. Hicks, and S. D. Andrews. In press. In-situ bioremediation: an integrated system approach. In Bioremediation: Field Experiences, P. E. Flathman, D. E. Jerger, and J. H. Exner, eds. Chelsea, Mich.: Lewis Publishers.

available and can be used in microcosm tests to trace the pollutant's fate very precisely. Comparing the microcosm-generated biodegradation rates under a variety of conditions can provide valuable information concerning whether environmental conditions in the field are conducive for high degradation rates. The careful control and monitoring possible in microcosms make rate determinations much less ambiguous than rates measured in the field.

Methods that rely on laboratory microcosms have uncertainties associated with directly extrapolating the laboratory results to the field. The delicate balance of chemical, physical, and biological relationships that influence bioremediation can change rapidly with environmental disturbances, such as to oxygen concentration, pH, and nutrient concentration. Research has shown that microbes removed to the laboratory may behave differently from those in the field—

quantitatively and qualitatively. Thus, laboratory experiments may impose artifacts that distort the interpretation of field conditions.

Bacterial Adaptation

Over time, bacteria at a contaminated site may develop the capability to metabolize contaminants that they were unable to transform—or that they transformed very slowly—when the contaminant was first spilled. Thus, metabolic adaptation provides evidence of bioremediation in the field. Adaptation can result from an increase in the number of bacteria able to metabolize the contaminant or from genetic or physiological changes within the individual bacteria.

Microcosm studies are well suited for assessing adaptation. An increase in the rate at which microorganisms in the sample transform the contaminant in microcosm tests provides evidence that adaptation has occurred and bioremediation is working. The rate increase can be determined by comparing samples from the bioremediation zone with samples from an adjacent location or by comparing rates before and after bioremediation.

Developments based on tools used in molecular biology may provide new methods for tracking whether bacteria have adapted to degrade certain contaminants. Gene probes specifically targeting degradative genes can be constructed and can, at least in principle, determine if that gene is present in a mixed population. Using probes in this manner requires knowledge of the DNA sequence in the degradative gene.

For the special case when a genetically engineered microorganism is applied to a site for bioaugmentation, the engineered organism can be fitted with a reporter gene that is expressed only when a degradative gene of interest also is expressed. Thus, the protein product of the reporter gene signals (for example, by emitting light) that the degradative gene is present *and* is being expressed in the in situ population.

Inorganic Carbon Concentration

In addition to more microbes, bacteria produce inorganic carbon—usually present as gaseous CO_2, dissolved CO_2, or HCO_3^-—when they degrade organic contaminants. Therefore, samples showing enrichment of the water and gas phases with inorganic carbon indicate active biodegradation. Gas chromatography is the method of choice for determining gaseous CO_2 concentrations; inorganic carbon analysis is appropriate for water samples.

Monitoring changes in inorganic carbon is inaccurate where high background bicarbonate concentrations or dissolution of calcareous minerals masks respiratory production of inorganic carbon. In these cases, stable isotope analysis of the carbon (described below) is a possible means to distinguish bacterially produced inorganic carbon from mineral carbon, but these techniques are still in the early research stages.

Carbon Isotope Ratios

One way of determining whether the CO_2 and other inorganic carbon in a sample is an end product of contaminant biodegradation or whether it originates from some other source is to analyze the sample's carbon isotopes. Most of the carbon will be present as the isotope ^{12}C (having six protons and six neutrons in its nucleus), but some will be present as ^{13}C (having six protons and seven neutrons in its nucleus, thus weighing slightly more than ^{12}C). The $^{13}C/^{12}C$ ratio of the inorganic carbon in a sample varies depending on where the carbon originated—from contaminant degradation, degradation of other organic matter, or mineral dissolution. Depending on the situation, $^{13}C/^{12}C$ ratios can be used in one of two ways.

The first use of $^{13}C/^{12}C$ ratios is appropriate when the carbon in the organic contaminant has a substantially different $^{13}C/^{12}C$ ratio than the inorganic carbon derived from mineral dissolution. This situation is relatively common because inorganic carbon from minerals contains substantially more ^{13}C than carbon derived from most organic contaminants. Although the $^{13}C/^{12}C$ ratio changes somewhat when organic contaminants are biodegraded to CO_2, inorganic carbon produced from most organic contaminants remains substantially more enriched in ^{12}C than inorganic carbon dissolved from mineral deposits. Thus, if the inorganic carbon taken from site samples has a $^{13}C/^{12}C$ ratio much lower than the ratio for carbon from mineral sources, it is likely that the carbon originates from contaminant biodegradation.

The second type of application exploits isotope fractionation, in which microbial metabolism usually creates inorganic carbon that is enriched in ^{12}C, while the remaining organic contaminant source becomes enriched in ^{13}C. For example, microorganisms degrade the lighter (^{12}C) isotopic forms of petroleum hydrocarbons more quickly than they degrade the heavier (^{13}C) forms. As a result, the petroleum hydrocarbons remaining in the subsurface become relatively enriched in ^{13}C as bioremediation proceeds. Thus, observation of a decreasing $^{13}C/^{12}C$ ratio in inorganic carbon, coupled with an increase in the ratio for the organic source, usually provides evidence that the inorganic carbon is being produced by contaminant biodegradation.

An exception to the typical trend of decreasing $^{13}C/^{12}C$ ratios in inorganic carbon occurs during methanogenesis, in which the end product of contaminant biodegradation is not CO_2, but methane (CH_4). Methanogenic organisms consume CO_2 by converting it to CH_4. In the process the pool of CO_2 becomes depleted in ^{12}C, while the methane generated by the organisms becomes enriched in ^{12}C. Thus, in methanogenic environments the $^{13}C/^{12}C$ ratio observed in samples of inorganic carbon may increase, instead of decreasing. Meanwhile, in the methane—the final sink for the carbon from the contaminant—the $^{13}C/^{12}C$ ratio decreases.

The $^{13}C/^{12}C$ ratio can be determined by analyzing samples with a mass spectrometer, a standard chemist's tool for separating isotopes and determining the relative masses of chemical compounds. The procedures for determining isotope ratios are elaborate, expensive, and only pertinent if the characteristic $^{13}C/^{12}C$ ratio of the contaminant source can be ascertained. Today, the $^{13}C/^{12}C$ ratio is an experimental method that requires further development and evaluation before it can be used as a definitive indicator of bioremediation. However, given the proper circumstances, the approach is advantageous because there is no requirement for sampling adjacent areas outside the bioremediation zone to evaluate relative responses (although the contaminant's characteristic $^{13}C/^{12}C$ signature must be determined from a sample representative of the source). Another potential advantage is that sampling for inorganic carbon does not require unusual precautions or equipment.

Electron Acceptor Concentration

In the process of transforming contaminants, bacteria consume electron acceptors, usually O_2, NO_3^-, or SO_4^{2-}, as explained in Chapter 2. A depletion in the electron acceptor concentration that occurs simultaneously with contaminant loss is further evidence that bioremediation is occurring. The electron acceptor concentration can be determined by standard analyses in wet chemistry. Sampling for O_2 must be carried out with extreme care to prevent increases in the sample's dissolved O_2 concentration due to contact with air.

Byproducts of Anaerobic Activity

Some of the key organisms useful in bioremediation are anaerobic—that is, able to exist without oxygen. These anaerobes are valuable because they are able to carry out many important biotransformation reactions when the supply of oxygen is limited. In addition, certain anaerobes are best able to carry out the initial dechlorination steps for highly chlorinated solvents and PCBs (see Table 2-1). Increases in metabolic products produced by anaerobes can signal an increase in anaerobic activity and indicate successful bioremediation (see Box 2-2). Key byproducts of anaerobic respiration include methane, sulfides, reduced forms of iron and manganese, and nitrogen gas. When significant amounts of chlorinated compounds are biotransformed, increases in chloride ion also may be observable. These measurements give the strongest evidence when parallel measurements confirm an anaerobic (oxygen-depleted) environment, loss of electron acceptors other than oxygen (for example, nitrate and sulfate), and consumption of electron donors responsible for the loss of the electron acceptors.

Intermediary Metabolite Formation

Microbiological processes may transform contaminants into unique intermediary metabolites. For example, during cometabolic microbial transformation of trichloroethylene, trans-dichloroethylene oxide may be produced. Detection of such metabolites from field samples provides evidence that in situ biodegradation is progressing (see Box 4-3 for an example). Intermediary metabolites can be determined by using gas chromatography, high-performance liquid chromatography, or one of these methods coupled to mass spectrometry. To ensure validity of this approach, the intermediary metabolites cannot have been present in the originally released contaminants, and they should be absent in adjacent uncontaminated areas. Because some intermediary metabolites degrade too quickly to be detected, an absence of intermediates does not indicate that bioremediation is not occurring.

Ratio of Nondegradable to Degradable Substances

If a site contains mixtures of contaminants, a decrease in the ratio of biodegradable to nonbiodegradable organic compounds over time can indicate microbiological activity in the field (see boxes 4-2 and 4-3). For example, phytane, a molecule that occurs in crude oil, is more resistant to microbial attack than octadecane, another crude oil component. Phytane and octadecane have the same molecular weight and similar volatility and transport characteristics and, consequently, are likely to undergo nearly identical abiotic reactions. Therefore, a decrease in the ratio of octadecane to phytane is evidence that microbes are degrading the octadecane. A possible drawback of this

BOX 4-3
TESTING BIOREMEDIATION OF PCBs IN HUDSON RIVER SEDIMENTS—NEW YORK

Laboratory studies have shown that polychlorinated biphenyls (PCBs)—substances once thought highly resistant to microbial attack—can, in fact, be biodegraded. In an effort to demonstrate the practical implications of these studies, the General Electric Corporation sponsored a 10-week in situ biodegradation field test. Researchers anchored six large adjacent cylinders in shallow areas of the Hudson River where the sediments are contaminated with lightly chlorinated PCBs. They tested the ability of native microbes and PCB-degrading microbes brought to the site from the laboratory to degrade the PCBs in place when stimulated with oxygen, a complete mixture of nutrients, and biphenyl to stimulate PCB cometabolism. While addition of the laboratory bacteria had no effect on in situ PCB degradation, significant destruction of PCBs occurred, and the researchers attributed the loss to biodegradation by the native microbes.

In demonstrating in situ bioremediation, the researchers provided the three key types of evidence outlined in this report:

1. *Documented loss of contaminants:* Over the 10-week test, between one-third and one-half of the PCBs were destroyed. The researchers determined PCB losses by measuring PCB concentrations in 12 cores in each cylinder at the beginning and end of the experiment.

2. *Laboratory assays showing that microorganisms have the potential to degrade the contaminants:* For this type of evidence, the researchers relied on several published laboratory studies showing that lightly chlorinated PCBs are susceptible to aerobic biodegradation.

3. *Evidence that biodegradation potential is realized in the field:* The most important evidence of in situ biodegradation was tests showing that chlorophenols—key intermediary metabolites in PCB degradation—appeared in the test cylinders after the microbes were supplied with the necessary nutrients, biphenyls, and oxygen. In addition, the researchers showed that the ratio of degradable to nondegradable PCBs decreased over time, indicating microbial attack of the degradable portions.

Reference

Harkness, M. R., J. B. McDermott, D. A. Abramowicz, J. J. Salvo, W. P. Flanagan, M. L. Stephens, F. J. Mondello, R. J. May, J. H. Lobos, K. M. Carroll, M. J. Brennan, A. A. Bracco, K. M. Fish, G. L. Warner, P. R. Wilson, D. K. Dietrich, D. T. Lin, C. B. Morgan, and W. L. Gately. 1993. In situ stimulation of aerobic PCB biodegradation in Hudson River sediments. Science 259(Jan. 22):503-507.

approach is that naturally occurring microbes often eventually attack phytane, causing the octadecane/phytane ratio to underestimate in situ biodegradation rates. Another example of this strategy is differential removal of volatile organic compounds that have roughly the same transport and volatilization properties but that are degraded at different rates. For instance, dichloroethane behaves nearly identically to trichloroethylene, but, unlike trichloroethylene, dichloroethane is not readily degraded under anaerobic conditions.

This approach is also useful for single contaminants having different forms, one of which is biodegradable and the other of which resists biodegradation. Some organic contaminants consist of mixtures of stereoisomers—molecules that are formed of the same elements and the same bonds but that have different spatial arrangements of the atoms. Hexachlorocyclohexane, for example, exists in two different forms, only one of which is readily metabolized. Thus, chemical analyses documenting selective disappearance of the degradable form of this contaminant are evidence of bioremediation. This approach is contaminant specific and requires substantial prior biochemical and physiological knowledge, but it illustrates an important principle that in the future could be of practical value in bioremediation projects.

Experiments Run in the Field

Several useful methods for evaluating whether microorganisms are actively degrading the contaminant involve not just sampling the site but also conducting active experiments in the field. These field experiments require adding various chemicals to the subsurface in a strictly controlled manner to see if their fate is consistent with what should occur during bioremediation.

Stimulating Bacteria Within Subsites

One type of field experiment involves adding materials that stimulate biodegradation to subsites within the contaminated area. Addition of stimulants such as electron acceptors, electron donors, and nutrients should speed biodegradation but not abiotic contaminant removal processes. Thus, when stimulants are added to one subsite but not another, the relative rate of contaminant loss should increase in the stimulant-amended subsites. The contrast in contaminant loss between enhanced and unenhanced subsites can be attributed to microbial activity. Applying this approach requires a setting uniform enough to have comparable subsites.

Measuring the Electron Acceptor Uptake Rate

A second field experiment involves alternately starting and stopping the supply of oxygen or other electron acceptors to the site to determine the rate at which the electron acceptors are consumed. This approach is particularly useful with air sparging because the oxygen supply can be controlled rapidly and independently of water flow. Immediately after stopping the flow of sparged gases, an oxygen probe is lowered into ground water wells to measure the rate of oxygen consumption. To distinguish oxygen used by contaminant-degrading microbes from oxygen used by ordinary microbial activity, background oxygen uptake rates should be measured in adjacent uncontaminated wells. Relatively rapid oxygen loss in the contaminated area compared to the uncontaminated area, coupled with a drop in the contaminant concentration, suggests successful bioremediation.

Monitoring Conservative Tracers

A third type of field experiment requires adding a *conservative tracer* to the site. Conservative tracers have chemical and transport properties similar to those of microbiologically reactive chemicals but are not microbiologically reactive themselves. Thus, conservative tracers can be used to distinguish abiotic chemical changes—such as volatilization, sorption, and dilution—from chemical changes caused by microorganisms.

One possible use of conservative tracers is to determine how much sparged oxygen is being consumed by microbes and how much is disappearing through abiotic routes, such as dilution. For this determination, helium gas (He) can be used as a conservative tracer for O_2. A known concentration of He is incorporated into the sparging system used to supply O_2 to the contaminated zone, and the changing concentrations of both He and O_2 are measured over time using a portable gas chromatograph and oxygen meter or other appropriate instruments. The rate of O_2 depletion relative to He depletion indicates the rate at which microbes are consuming O_2. If O_2 is being consumed at a rate related to the contaminant consumption rate, it is likely that microorganisms are responsible for contaminant disappearance. In some cases, O_2 can be consumed abiotically, such as by iron oxidation (converting Fe^{2+} to Fe^{3+}). When such a possibility exists, O_2 depletion measured in comparison with a tracer should also be determined in background uncontaminated zones to estimate the abiotic O_2 consumption rate.

For sites where a dissolved chemical (such as NO_3^-, SO_4^{2-}, or dissolved O_2) is the electron acceptor, bromide can be used as a conservative tracer. In this approach the bromide is added to the water circulated through the ground to supply the electron acceptor.

Although conservative tracers that mimic contaminant behavior often are added to the site, they also may be fortuitously present in the contaminant. As discussed under "Measurements of Field Samples," some contaminants contain mixtures of degradable and nonbiodegradable compounds that move through the subsurface in similar ways. When the concentration of degradable compounds drops faster than the concentration of conservative tracers, the difference can be attributed to microbial activity in the field.

Labeling Contaminants

A fourth type of field experiment involves monitoring the fate of "labeled" contaminants. Contaminants can be labeled by synthesizing versions in which the contaminant molecules contain a known amount of a stable isotope, usually ^{13}C or deuterium (a hydrogen isotope). If the expected metabolic byproducts, such as inorganic carbon and intermediary metabolites, carry the same relative amounts of ^{13}C and deuterium as the labeled contaminants, bioremediation is occurring. This technique is useful primarily for field research and not commercial bioremediation because it involves synthesizing a special version of the contaminant, which is costly, and adding it to the site, which temporarily increases the level of contamination. In addition, contaminant labeling is useful only for situations in which the contaminant source can be located. Adding the labeled compound to the wrong location may result in a false negative.

Modeling Experiments

A final type of technique for evaluating whether bioremediation is occurring in the field uses models—sets of mathematical equations that quantify the contaminant's fate. Models keep track of all the contaminant mass that enters the subsurface, describing how much dissolves, how much sorbs to solids, how much reacts with other chemicals, how much flushes out in the water, and how much biodegrades. The goal of using models is to see whether predictions of contaminant fate based on interpretation of the phenomena taking place during the bioremediation, as described by the model, match what is happening in the field, as determined by field sampling.

Contaminated field sites can be efficiently managed with the aid of models because models provide a means for synthesizing all relevant information. Furthermore, because models quantitatively link many types of measurements, they assist in evaluating the significance of a limited number of field observations. When models are sufficiently accurate, they may be powerful tools for assessing bioremediation.

Depending on the type and amount of data, the stage of process understanding, and the types of questions being asked, models can vary from very simple to highly complex. For example, a conceptual model, which does not yet have mathematical equations, may be appropriate when limited data are available during initial site characterization. On the other hand, complex mathematical models, solved on high-speed computers, become possible and more appropriate as understanding of the site expands during design and operation of a bioremediation project.

Types of Models

Because so many complex processes interact in the subsurface, four different types of models have been developed: saturated flow, multiphase flow, geochemical, and reaction rate models. Each model describes a different suite of subsurface processes and is used in particular ways to evaluate bioremediation. Ultimately, researchers often combine two or more types of models to do a complete evaluation.

Saturated Flow Models. Saturated flow models start by describing where and how fast the water flows through the saturated zone (the region below the water table). These models are derived from basic principles of conservation of fluid mass. Saturated flow is reasonably well understood, and the basic forms of these models for water flow are relatively simple, accurate, and accepted.

Once the direction and velocity of water flow are known, saturated flow models can be extended to describe the movement of dissolved contaminants. These contaminant transport models are based on principles of conservation of chemical mass. When the model contains no terms for reactions, it describes the fate of a conservative tracer. The conservative material basically moves with the water flow, although it is subject to processes that disperse, or mix, the contaminants.

Sorption of contaminants to the solid matrix slows the movement of the dissolved contaminants, compared to the water. Sorption effects often can be modeled simply by incorporating "retardation factors" that reflect the slower rate of transport of the contaminant relative to the water. In other cases, sorption phenomena are more complex than can be captured by a simple retardation factor and must be modeled using equations that consider sorption and desorption rates.

In special cases, biodegradation reactions can be described by very simple expressions (for example, first-order decay) that are easily incorporated into the transport part of a saturated flow model. However, many biodegradation phenomena are too complex to be incorporated so simply into a saturated flow model. Special modeling tools are needed and are discussed in the section below on biological reaction rate models.

Multiphase Flow Models. Whereas saturated flow models describe the flow of only one fluid, the ground water, through a porous medium, multiphase flow models describe the situation in which two or more fluids exist together in the porous medium. The fluids can be liquids or gases. The most common multiphase flow models predict the movement of water and contaminants above the water table, where a gas phase is present. This situation is called unsaturated flow. Addition of a light nonaqueous-phase liquid contaminant such as gasoline, which resides at or near the top of the water table, is a further complication that may be considered in models of unsaturated flow. Multiphase flow models also can describe the flow of dense nonaqueous-phase liquids such as chlorinated solvents, which move in a distinct mass separate from the ground water.

The phenomena controlling multiphase flow are not as well understood and are much more difficult to represent mathematically than are those for water flow in the saturated zone because they involve complex interactions among solids, water, air, and nonaqueous phases. The accuracy of multiphase flow models for water direction and velocity is limited by the large number of required transport parameters. Furthermore, the modeling community has not yet reached a consensus as to which modeling approach is most valid. Despite these limitations, multiphase flow modeling provides a framework for conceptualizing the movement of fluids in the subsurface and for making order-of-magnitude estimates of fluid movement.

If the direction and velocity of fluid flows can be predicted, modeling contaminant transport with multiphase flow models is similar to that for saturated flow. However, contaminant transport is complicated by the multiple phases, which introduce heterogeneities that affect dispersion and sorption.

Geochemical Models. At many contaminated sites, the contaminants are subject to a significant number of different chemical reactions.

Geochemical models describe how a contaminant's chemical speciation is controlled by the thermodynamics of the many types of chemical reactions that may occur in the subsurface. Today, geochemical models are used primarily to understand the fate of inorganic compounds. For example, these models can be used to analyze the series of reactions that influence whether a particular metal will precipitate. Geochemical models also can aid in determining the availability (solubility) of nutrients and trace metals required for microbial metabolism.

Although they are valuable, geochemical models have had limited use for assessing bioremediation. There are three reasons for the relatively low level of use. First, existing commercial applications of bioremediation have focused on aerobic biodegradation of petroleum hydrocarbons, a situation for which inorganic geochemistry usually is not a crucial factor. As bioremediation is applied to more complex sites, especially those containing contamination by heavy metals, the need for geochemical modeling will increase. Second, traditional geochemical models are founded on the principle of equilibrium conditions—in other words, all possible reactions are assumed to occur to their maximum possible extent. The equilibrium assumption typically is not valid for bioremediation because the key reactions are almost always controlled by kinetics—the rate at which a reaction moves toward equilibrium. Third, traditional geochemical models are very complicated and expensive to use, even when they are not connected to transport modeling. Therefore, their use has been limited to evaluating possible changes in subsurface chemicals.

Biological Reaction Rate Models. Biological reaction rate models represent how quickly the microorganisms transform contaminants. They are useful for evaluating bioremediation systems because the rate at which the microbes work is the key factor influencing how much time the cleanup will take.

The rate of biodegradation depends on the amount of active biomass present; the concentrations of contaminants, electron acceptors, and other "food" sources for the bacteria; and certain parameters that describe transport rates of key chemicals to the bacteria and rates of enzyme-catalyzed reactions. All of this information can be packaged into a rate expression of the form:

$$\text{rate of biotransformation} = q_{max}\, X\;\; f(S_1, S_2, \ldots.)$$

in which q_{max} describes the reaction rate per unit amount of biomass for optimal conditions, X is the amount of active biomass, $f(S_1, S_2, \ldots.)$ is a mathematical function that describes how substrate transport and concentration reduce the rate from the optimal rate, and $S_1,S_2,....$ represent different substrates that participate in the reaction. The value of X is not necessarily constant; it can change with time and location. Keeping track of X is part of the model. The $f(S_1,S_2,....)$ function can range from very simple, such as the concentration of just one substrate, to complex sets of equations involving several substrates and rate parameters. Determination of appropriate rate expressions and parameter values for those expressions is an active research area.

Combining Models. In many cases, evaluating bioremediation involves combining two or more of the model types. For example, in situ bioremediation of a chlorinated solvent may require a multiphase flow model coupled to a sophisticated biodegradation rate model. The multiphase flow model tracks the movements of the water and the solvent; once the flows are known, a transport model uses a biodegradation rate model as a sink term.

Biodegradation rate models are most easily combined with flow models when one rate-limiting material can be identified. The rate-limiting material often is the primary electron donor or electron acceptor. For example, the biodegradation rate of petroleum hydrocarbons often can be modeled with dissolved oxygen as the rate-limiting substance. In several successful modeling studies, overall biodegradation rates could be modeled by the rate at which oxygen entered the bioremediation zone.

The major simplification achieved by assuming rate limitation solely by oxygen should not be considered a general rule. It can be appropriate for biodegradation of petroleum hydrocarbons (a process that is especially sensitive to low oxygen concentrations) when the input rate for dissolved oxygen is low compared to the amount of hydrocarbon present and the site is large. Because these conditions are not true in many other situations, biodegradation rate modeling may require different and more sophisticated approaches.

Except when the biodegradation or geochemical models are very simple, coupling them with flow models requires more than an extension of the existing contaminant transport models used for conservative tracers. Considerable attention must be given to proper model formulation and to efficient and accurate solution techniques. Otherwise, costs and computer time will be excessive.

How to Use Models

Models provide a framework for organizing information about

contaminated sites. They increase understanding of contaminant behavior by requiring the model user to confront details such as the mass of contaminants, their chemical properties, and their dynamic interactions with site hydrogeochemical characteristics. When this required information is available and integrated into the proper model, modeling predictions become useful tools for managing field sites and evaluating bioremediation.

Models can be useful for evaluating in situ bioremediation in two ways. One approach is to see if a model representing only abiotic mechanisms can or cannot account for all of the contaminant loss. A second approach goes a step further and evaluates if "reasonable" estimates of microbial processes, quantified through the model, can explain contaminant losses (see Box 4-4). This second approach requires detailed knowledge of rate coefficients describing how quickly the microbes degrade the contaminant, in addition to parameters describing transport and other abiotic phenomena.

Mass Losses. The first modeling approach requires analyzing whether abiotic mechanisms (for example, dilution, transport, and volatilization) can explain all of the losses of the contaminant mass. The approach recognizes that biodegradation rate models often have greater uncertainty that do models of abiotic processes. The uncertainty can be caused by poor understanding of the biochemical reactions, difficulty estimating parameters, and inadequate site characterization. Eliminating biological reactions from the model avoids this uncertainty.

When the model of abiotic mass losses shows that some contaminant mass remains after all the abiotic sinks are considered, there are two possible explanations: (1) biodegradation processes are implicated as the sink for the "missing" mass, or (2) the modeling parameters were improperly selected, have led to inaccurate predictions, and are therefore misleading the modeler. Because judgments about microbiological involvement in contaminant loss may be contingent upon the selection of parameters used to describe abiotic losses, a modeler must be vigilant—constantly scrutinizing the validity of decisions and parameters that affect the modeling results. Adjustments in modeling parameters can lead to vastly different predictions; therefore, it is prudent to give credence to evidence for bioremediation only when the modelers have a high degree of confidence in their results and when discrepancies between actual and modeled contaminant behavior are unambiguous. Conclusions about effective bioremediation should only be drawn when concentrations of contaminants found in field sites are not simply lower but *significantly* lower than would be

BOX 4-4
PROVING INTRINSIC BIOREMEDIATION OF A SPILL AT A NATURAL GAS MANUFACTURING PLANT—NORTHERN MICHIGAN

At a plant in northern Michigan, waste products from natural gas manufacturing leaked from a disposal pit into the surrounding ground water. Having installed wells around the plant to prevent off-site migration of contaminated water, the company in charge of the facility chose intrinsic bioremediation to clean up the contaminants (primarily benzene, toluene, and xylene, or BTX). In demonstrating the effectiveness of bioremediation, the company provided evidence that meets the three criteria discussed in this report:

1. *Documented loss of contaminants:* The company began its extensive site-monitoring program to follow the effectiveness of intrinsic bioremediation in 1987. Since that time the benzene concentration has dropped by approximately 90 percent and the contaminant plume has shrunk considerably.
2. *Laboratory assays showing that microorganisms have the potential to degrade the contaminants:* The company performed a series of lab tests with soil cores retrieved from the field showing that the site's native microbes could degrade BTX at a high rate—5 to 10 percent per day—if supplied with adequate oxygen (1 to 2 ppm or more).
3. *Evidence showing that biodegradation potential is realized in the field:* The company used a computer-based model, BIOPLUME II, to demonstrate that the rate of contaminant loss that one would predict if bioremediation were occurring closely matched the actual contaminant loss rate in the field. In 1987 the company measured the BTX and dissolved oxygen levels at various points in the plume. These values were input into BIOPLUME II to predict how they should change with time if bioremediation were occurring. The field measurements of both the contaminant concentrations and the dissolved oxygen levels taken since 1987 closely match the model's predictions. In addition, the biodegradation rate predicted by the model closely matches the rate measured in the field.

Monitoring at this site is still ongoing to demonstrate the long-term effectiveness of intrinsic bioremediation.

Reference

Chiang, C. Y., J. P. Salanitro, E. Y. Chai, J. D. Colthart, and C. L. Klein. 1989. Aerobic biodegradation of benzene, toluene, and xylene in a sandy aquifer: data analysis and computer modeling. Groundwater 27(6):823-834.

expected from predictions based on abiotic processes. Thus, this approach works well when biodegradation is the dominant sink and when the abiotic processes are well characterized. Uncertainty in modeling the abiotic processes makes this approach unreliable when biodegradation is not the dominant removal mechanism.

Direct Modeling. When reasonable estimates of biological processes and parameters are available, directly modeling the biodegradation process is the superior modeling strategy. These estimates can be obtained from the scientific literature, past experience with similar circumstances, laboratory experiments, or field-scale pilot studies, depending on the site conditions and biodegradation reactions.

One approach is to use the model to answer the question, "Does our best representation of the biodegradation rates, when combined with the simultaneously occurring abiotic rates, support the conclusion that biological reactions are responsible for observed changes in contaminant levels or other relevant observations?" If the answer is "yes," modeling provides a much greater measure of confidence that observations supporting biodegradation are not artifacts.

A second approach is to use direct modeling to predict the contaminant's concentration at unsampled locations or to predict the future concentration. The model then identifies sample locations and times that should yield particularly definitive measurements. Subsequent sampling, if consistent with model predictions, confirms the analyst's understanding of what is occurring in the subsurface. Lack of agreement between model predictions and actual developments forces a reevaluation of the model and improves understanding of the site and the parameter values.

In some cases, direct modeling must involve highly sophisticated computer codes that take into account the three-dimensional nature of the site, heterogeneities, and complex reactions. These models are expensive to formulate and run, but they are essential tools for investigators who require a detailed description of what is happening at a site. Currently, these types of models are viewed primarily as research tools appropriate for highly monitored research, demonstration, or pilot sites.

In many practical applications, direct modeling can be greatly simplified by eliminating all but the most essential phenomena. A good strategy is to compare expected rates of all phenomena that might affect the bioremediation. For example, the expected rate of contaminant loss due to biodegradation can be compared with the expected contaminant loss rate due to volatilization. Normally, a few of the possible phenomena will have expected rates much greater than those of the other phenomena, and the model can consider only the phenomena having relatively high rates. If the biodegradation rate is high enough that it should remain in the model, the model provides *prima facie* evidence that bioremediation is working. Solution of the complete model can verify the evidence.

Limitations of Models

Although a powerful tool, modeling has its shortcomings. One shortcoming is that a model's validity must be established on a site-by-site basis, because no "off-the-shelf" models are available for evaluating bioremediation on a routine basis. Although a drawback in terms of time and cost, model validation may be a net advantage because it results in a more complete understanding of the site. Another limitation is that determining each of the many modeling parameters (such as hydraulic conductivity, retardation factors, and biodegradation rate parameters) may be as demanding and expensive as making the measurements for other types of verification criteria. Thus, a trade-off may exist between better modeling and more field measurements.

Despite its limitations, modeling should be a routinely used tool for understanding the dynamic changes that occur in field sites during bioremediation. Although the complexity and type of model can vary, modeling is a valuable tool for linking conceptual understanding of the bioremediation process with field observations and for giving weight to a limited set of data. Even if site complexities preclude assembling a model that provides valid quantitative predictions, models are valuable management tools because they integrate many types of information relevant to the fate of contaminants.

LIMITATIONS INHERENT IN EVALUATING IN SITU BIOREMEDIATION

Because the subsurface is complex and incompletely accessible, knowledge of the fate of ground water contaminants always will be limited. This situation is intensified for in situ remediation technologies of any type, because frequently the amount, location, and type of contamination are unknown. Without knowing the starting point for a remediation, defining the finishing point is difficult. Errors in measurements, artifacts imposed by extrapolating lab results to the field, and an inherent shortage of data further complicate the evaluation and create uncertainty about the performance of a remediation process. For example, in analyzing chemical concentrations in ground water, a large number of samples from spatially different locations may be gathered. Even assuming the laboratory results are completely error free, uncertainty arises from extrapolating these point samples in an attempt to portray a complete picture of how the water's chemical composition varies in space.

Because evaluation of bioremediation requires integrating concepts and tools from very different disciplines, efforts to synthesize information from these different disciplines can create problems. For example, microbiologists and hydrogeologists use space and temporal scales that seldom match. The seconds and micrometers characteristic of microbial processes are very much smaller than the months and kilometers typical of hydrogeological descriptions of landscape processes. Thus, the hydrologic data describing large-scale water flow do not always meet a microbiologist's needs for understanding the small-scale mechanisms that control microbial activity. For instance, models efficient for the typical space scale of water movement (i.e., meters to kilometers) obliterate all of the details of microbial reactions, which occur in distances of micrometers to centimeters.

A prime example of the problem of trying to synthesize different scales is illustrated by the problems encountered when trying to document major increases in biomass during in situ bioremediation. Microorganisms often are highly localized near their food sources. This localization makes it difficult to "find" the organisms when only a few samples can be taken. Microbial numbers, biodegradation rate estimates, or biodegradation potentials can vary tremendously, depending on whether the sample was from a location of high microbial activity or from a nearby location with low activity. Microbiological variability occurs on a small scale compared to the scale represented by field samples. Consequently, uncertainty in microbiological parameters always is a risk.

Three strategies can help minimize uncertainty and should play important roles in evaluating bioremediation: (1) increasing the number of samples, (2) using models so that important variables are properly weighted and variables with little influence are eliminated, and (3) compensating for uncertainties by building safety factors into the design of engineering systems. Investigators can trade off these three strategies. For example, if gathering a large number of samples or using sophisticated models is not possible, larger safety factors can cover the resulting uncertainty. At small field research sites designed to investigate bioremediation of contaminants not yet treated on a commercial scale, a large number of samples and complex models may be possible—and necessary—to draw detailed conclusions from the research results. On the other hand, at large commercial sites, a similarly high density of samples may be cost prohibitive, and it may be more appropriate to rely on larger safety factors to account for the greater uncertainties.

Uncertainties in evaluating bioremediation can be minimized but not eliminated. Investigators cannot fully understand the details of whether and how bioremediation is occurring at a site. The goal in evaluating in situ bioremediation is to assess whether the weight of evidence from tests such as those described above documents a convincing case for successful bioremediation.

5

Future Prospects for Bioremediation

In preparing this report the National Research Council's Committee on In Situ Bioremediation sought to communicate the scientific and technological bases for bioremediation. As the report has explained, the principle underlying bioremediation is that microorganisms (mainly bacteria) can be used to destroy hazardous contaminants or transform them into less harmful forms. Microorganisms are capable of performing almost any detoxification reaction. Nevertheless, the commercial practice of bioremediation today focuses primarily on cleaning up petroleum hydrocarbons. The full potential of bioremediation to treat a wide range of compounds cannot be realized as long as its use is clouded by controversy over what it does and how well it works. By providing guidance on how to evaluate bioremediation, the committee hopes this report will eliminate the mystery that shrouds this highly multidisciplinary technology and pave the way for further technological advances.

This chapter summarizes new research advances that the committee foresees as expanding the future capabilities of bioremediation. It recommends steps that will improve the ability to evaluate bioremediation technologies objectively, whether the technologies are new or established.

NEW FRONTIERS IN BIOREMEDIATION

Bioremediation integrates the tools of many disciplines. As each of the disciplines advances and as new cleanup needs arise, opportunities for new bioremediation techniques will emerge.

Until now, three types of limitations have restricted the use of bioremediation to clean up contaminants other than petroleum hydrocarbons: inadequate understanding of how microbes behave in the field, difficulty supplying the microbes with stimulating materials, and problems with ensuring adequate contact between the microbes and the contaminant. Consequently, only a few of the myriad microbial processes that could be used in bioremediation are applied in practice. Recent advances in science and engineering show promise for overcoming these limitations, as illustrated by the following examples:

- **Understanding microbial processes**. As novel biotransformations become better understood at ecological, biochemical, and genetic levels, new strategies will become available for bioremediation. A recent example is microbial dechlorination of polychlorinated biphenyls (PCBs), which is being investigated by a group of researchers from academia and industry. The researchers, studying PCBs in Hudson River sediments, have documented that anaerobic microbes in the sediments can transform highly chlorinated PCBs to lightly chlorinated PCBs, which can be degraded completely by aerobic microbes (see Box 4-3). This research may become the basis for commercial bioremediation of PCBs—compounds once thought to be undegradable. Similar advances are being made for the dechlorination of chlorinated solvents, also once believed to resist biodegradation.

Advances in understanding microorganisms may also improve bioremediation's effectiveness in meeting cleanup standards. As explained in Chapter 2, uptakc and metabolism of organic compounds sometimes stop at concentrations above cleanup standards. Research on bioaugmentation and direct control of the cell's genetic capability and/or regulation is very active today and may lead to methods to overcome such microbiological limitations.

- **Supplying stimulating materials**. Innovative engineering techniques for supplying materials that stimulate microorganisms are pushing the boundaries of bioremediation. For instance, the recent innovation of gas sparging has substantially expanded capabilities for aerobically degrading petroleum hydrocarbons. Research is ongoing into optimizing ways to supply materials other than oxygen. Such research will pave the way for emerging bioremediation applications, such as degradation of PCBs and chlorinated solvents and demobilization of metals, which are not necessarily controlled by oxygen.

- **Promoting contact between contaminants and microbes**. Research is under way into engineering advances to increase the availability of contaminants to microbes—advances that, if successfully applied, would increase bioremediation's efficiency. New techniques for promoting contaminant transport to the organisms include high-pressure fracturing of the subsurface matrix, solubilization of the contaminants by injecting heat (via steam, hot water, or hot air), and, perhaps, addition of surfactants. Discovery of improved methods for dispersing the microorganisms may also enhance microbial contact with the contaminants and lead to more effective bioremediation.

THE INCREASING IMPORTANCE OF EVALUATING BIOREMEDIATION

As new bioremediation techniques are brought from the lab into commercial practice, the importance of sound methods for evaluating bioremediation will increase. The Committee on In Situ Bioremediation has recommended a three-part strategy for "proving" that bioremediation has worked in the field. As explained in Chapter 4, the three central parts of this strategy are (1) documented loss of contaminants from the site, (2) laboratory assays showing that microorganisms from site samples have the *potential* to transform the contaminants, and (3) one or more pieces of information showing that the biodegradation potential is *actually realized* in the field. The main goal of this strategy is to show that biodegradation reactions that are theoretically possible are actually occurring in the field, at fast enough rates and in appropriate locations to ensure that cleanup goals are met.

While the three-part strategy provides a general framework for evaluating bioremediation, the level of detail with which it should be applied depends on the interests of those involved with the bioremediation. Each party involved must realize that "success" may mean different things to the different parties. Regulators are primarily concerned that legislated standards are achieved, clients emphasize attaining cost-effective goals, and vendors have a vested interest in demonstrating that their technology is effective and predictable. Clear communications about everyone's goals and negotiations about specific criteria to meet the different goals are critical to the project's perceived success and must occur in advance of its implementation.

The current knowledge base is sufficient to allow implementation of the three-part strategy. However, the specific experimental protocols for carrying out the strategy need to be developed. In addition, further research and better education of those involved in bioremediation will improve the ability to implement the strategy as well as understanding of the fundamentals of bioremediation.

Recommended Steps in Research

The committee recommends research in the following areas to improve evaluations of bioremediation:

- **Evaluation protocols**. Protocols need to be developed for putting the three-part evaluation strategy into practice. Consideration should be given to evaluating a range of chemical contaminants (including petroleum hydrocarbons, chlorinated solvents, PCBs, and metals) and site characteristics (such as shallow and deep aquifers and sites with high and low heterogeneity). These protocols should be field tested through coordinated efforts involving government, industry, and academia and should be subject to scientific and peer review.
- **Innovative site characterization techniques**. Rapid, reliable, and inexpensive site characterization techniques would have a significant impact on the ease of evaluating bioremediation. Examples of relevant site measurements include distribution of hydraulic conductivities, contaminant concentrations associated with solid or other nonaqueous phases, native biodegradation potential, and abundance of different microbial populations. Techniques to measure physicochemical characteristics in situ are being developed and could revolutionize the capability to do field assessments. Methods adapted from molecular biology seem especially promising for augmenting current techniques for assaying biodegradation potential and microbial populations. Gathering more and better characterization data would diminish uncertainties and reduce the needs for overdesign via safety factors.
- **Improved models**. Improvements in mathematical models are essential because models link understanding of chemical, physical, and biological phenomena. One particularly promising advancement is the use of modeling as a key part or improved means for on-site management, which requires an appreciation of the dynamic interactions among the many phenomena. As field sampling becomes more rapid and accurate, on-site decisions will be limited more by the ability to understand the dynamic interactions than by turnover times between sampling and analysis.

Recommended Steps in Education

Steps need to be taken to educate all components of society about what bioremediation is and what it can and cannot do. Especially important is improved education of the people who are in direct decision-making positions. The committee recommends three types of education:

- **Training courses that selectively extend the knowledge bases of the technical personnel currently dealing with the uses or potential uses of in situ bioremediation**. This step explicitly recognizes that practitioners and regulators who already are dealing with complicated applications of bioremediation need immediate education about technical areas outside their normal expertise.
- **Formal education programs that integrate the principles and practices for the next generation of technical personnel**. This step explicitly recognizes the need to educate a new generation of technical personnel with far more interdisciplinary training than is currently available in most programs.
- **Means for effective transfer of information among the different stakeholders involved in a project**. Effective transfer requires that all types of stakeholders participate, that all are invested in achieving a common product (e.g., a design, a report, or an evaluation procedure), and that sufficient time is allocated for sharing perceptions and achieving the product. This step may involve more time and more intensive interactions than have been the norm in the past.

In summary, in situ bioremediation is a technology whose full potential has not yet been realized. As the limitations of conventional ground water and soil cleanup technologies become more apparent, research into alternative cleanup technologies will intensify. Bioremediation is an especially attractive alternative because it is potentially less costly than conventional cleanup methods, it shows promise for reaching cleanup goals more quickly, and it results in less transfer of contaminants to other media. However, bioremediation presents a unique technological challenge. The combination of the intricacies of microbial processes and the physical challenge of monitoring both microorganisms and contaminants in the subsurface makes bioremediation difficult to understand—and makes some regulators and clients hesitant to trust it as an appropriate cleanup strategy. The inherent complexity involved in performing bioremediation in situ means that special attention must be given to evaluating the success of a project. Whether a bioremediation project is intrinsic—

relying on the natural properties of the subsurface—or engineered—augmenting subsurface properties to promote microbial activities—the importance of a sound strategy for evaluating bioremediation will increase in the future as the search for improved cleanup technologies accelerates.

Background Papers

A Regulator's Perspective on In Situ Bioremediation

John M. Shauver
Michigan Department of Natural Resources
Lansing, Michigan

SUMMARY

Bioremediation, like any technology applied to clean up a contaminated site, must first be approved by government regulators who ultimately must agree that the technology has a reasonable chance to reduce the contaminant(s) at the site to acceptable levels. This paper describes the information that regulators need to make their decision. Basically, this information comprises descriptions of the site, the specific cleanup process, and the overall approach to site cleanup. The paper also answers the questions of who, what, when, where, and how in the context of bioremediation on the basis of my 24 years of experience as a regulator.

INTRODUCTION

During the past 20 years, various companies and individuals have developed or claim to have developed biological treatment processes that could clean up various wastes generated by human activities. These wastes include polychlorinated biphenyls (PCBs), crude oil, refined crude oil products, crude oil wastes, and DDT, to name a few. One problem that the proponents of such treatment technology face is state and federal regulations. It is often hard for the regulated community to understand what is required to ensure that the regulator will approve a proposed treatment process.

This paper describes what a regulator looks for in a proposal to clean up (remediate) a site to legal standards. The guidance provided here is a condensation of the requirements of the many statutes and regulations used by the Michigan Department of Natural Resources. The paper reflects my view, after 24 years as a regulator, of what information needs to be routinely provided to evaluate a cleanup technology before it is applied to a particular site. Complex sites with unique or unusual features may have to be characterized in greater detail before a cleanup technology can be chosen. Also, the regulated community (potentially responsible parties) must realize that the cleanup process itself is but one facet of the overall site cleanup. To gain approval for implementation of a cleanup process, the responsible party should supply information that includes:

- a description in three dimensions of the site and of the type and extent of contamination,
- a detailed description of the cleanup process(es) to be applied to the site, and
- a detailed description of the approach to overall site cleanup.

SITE DESCRIPTION

The site description should specifically identify the types and amounts of chemical(s) released to the soil and ground water and other phases of the site environment. The description should also include estimates of the rate of movement of the contaminants through the various phases of the environment and of where they are likely to end up. The regulator's response to a given situation depends strongly on the rate of transport and the likely fate of the contaminants.

The site is the three-dimensional area contaminated by the chemicals that have been released. The site is not limited to legal property boundaries. In fact, it usually involves more than one property owner, and the owners may not all be responsible for the contamination. The site description should also include the vertical, horizontal, and lateral extent of contamination, which includes:

- soil type(s), permeability, porosity of the soils and/or aquifer, and concentrations of contaminants in soil;
- if appropriate, depth to ground water, rate and direction of flow, concentrations of contaminants in ground water, and concentrations of naturally occurring or other compounds (inorganic or organic) that may interfere with the treatment process;
- if appropriate, concentrations of contaminants in the air, prevailing wind direction, and nearest human receptors; and

- concentrations of contaminants in surface waters and sediments.

In any site description the regulator will place great emphasis on identifying the location of the source(s) of contamination. Removal of these sources, or hot spots (identified by an adequate site investigation), is the most effective way to limit migration of chemicals off site. In addition, elimination of the source of the contamination as early as possible is one of the most cost-effective ways to limit future cleanup costs.

A site description should also describe the process that caused the release. This is important because the regulator is required to determine the full extent of the type of contamination at the site. If the material released is gasoline, for example, it is very important to know whether it is leaded or unleaded and whether it came from a hole in a tank; an overfilled tank; or faulty pipes, valves, or other fittings. If the release is described as crude oil, it is important to know if brine, condensate, or other materials are present as well. The description of the cause of the release allows the regulator to identify its source and thus the most highly contaminated areas of the site.

PROCESS DESCRIPTION

The responsible party should provide a detailed description of the treatment process to be used. The engineer who is accustomed to describing an activated carbon process should provide the same level of detail for a biological process. The description should show how the process chosen will contain, destroy, or remove the contaminants to meet legal standards. If biological treatment is chosen, the regulator must be given data that show the ability of the organisms present in or added to the contaminated area to safely and effectively treat the chemical(s) on the site.

When living organisms are proposed to clean up a site, the regulator expects to see a detailed description of the organisms' requirements for oxygen, nutrients, temperature, moisture, and pH. We must be sure the organism will thrive long enough to treat the chemicals to legal cleanup standards. In addition, if an anaerobic treatment scenario (such as one using iron or sulfur) is proposed, the regulator needs to know that native microbes are capable of the proposed metabolism and that ambient or added nutrients will be available in amounts likely to allow effective treatment but not likely to cause rapid plugging of the delivery wells and/or the soils.

We must be able to determine that the use of bacteria in the soils and ground water (if unsuccessful) will not prevent other treatment technologies from being applied. Use of organisms without adequate information or controls in the past has resulted in severe plugging problems in ground water monitoring wells and/or the aquifer itself. Such loss of permeability not only prevents delivery of the nutrients and oxygen necessary to sustain biological activity to clean up the soils or aquifer but may seriously hamper use of other technologies.

OVERALL SITE CLEANUP DESCRIPTION

A very important part of the description of the overall approach to site cleanup is the method(s) to be used to prevent movement of the contaminants farther off site through the soil or to or through the ground water or other medium. Containment to prevent further spread of the chemicals is as important in the regulator's mind as any other part of the cleanup. The regulator needs a complete description of the steps to be taken to prevent further movement of the chemicals through the soil, air, ground water, or surface water.

For example, contaminated ground water may be moving downgradient at 15 cm per day. Purge and capture wells would have to be installed to pump this contaminated ground water back upgradient to the treatment system to prevent further movement of the contaminants off site. If the water is discharged to the ground surface via an infiltration bed, and if it contains volatile organic chemicals (VOCs) that would be released, the responsible party needs to demonstrate adequate control of VOC discharge to the air.

The description also should cover equipment necessary to achieve the cleanup. With biological treatment systems, equipment may be needed for adjustment of the pH of the ground water, removal of iron or other interfering substances before treatment, oxygen/air delivery or oxygen reduction, and identification and monitoring of tracers and nutrients added. For example, if the proposal is to use aerobic bacterial decomposition of the contaminant(s) and the contamination exists to a depth of 15 m below ground water surface in soils with a permeability of 10^{-7}cm/s, the regulator will be interested in how the responsible party intends to deliver oxygen or air and related nutrients to the organisms.

Also necessary is a description of the monitoring procedures to be used to show that the cleanup system is operating properly. When using biological systems, the responsible party must show that the organisms are, in fact, doing the job. For example, if an aerobic process is used, the level of oxygen in and around the plume of contamination in the ground water will have to be monitored to ensure that the organisms have sufficient oxygen to decompose the chemicals in the ground water. This type of monitoring may be in addition to or in place of simply monitoring for the contaminant itself. In addition, if nutrients are added, they may also be contaminants and require monitoring. Nitrate, for example, is a chemical of concern that may have to be added to a biological treatment system as a nutrient or may be proposed as an electron acceptor in an anaerobic treatment process. In Michigan the drinking water supplies may not contain more than 5 mg/l of nitrate. Therefore, if nitrate is used, the regulatory agency will require that it be monitored in addition to other monitoring requirements.

CONCLUSION

A regulator looks for the data necessary to determine that a proposed treatment technology, if properly installed and operated, will reduce the contaminant concentrations in the soil and water to legally mandated limits. In this sense the use of biological treatment systems calls for the same level of investigation, demonstration of effectiveness, and monitoring as any conventional system.

An Industry's Perspective on Intrinsic Bioremediation

Joseph P. Salanitro
Shell Development Company
Houston, Texas

SUMMARY

Laboratory and field evidence is now sufficient to demonstrate that soil microorganisms in aquifers are responsible for a significant portion of the attenuation of aromatic compounds—benzene, toluene, ethylbenzene, and xylenes (BTEX)—from fuel spills to the subsurface environment. Most subsoils contain indigenous microbes that can biodegrade low levels of BTEX (ppb to low ppm), given enough dissolved oxygen in the ground water. With adequate site characterization, analysis, and monitoring, this type of intrinsic bioremediation can shrink plumes and control the migration of hydrocarbons. In situ biodegradation processes, properly monitored, should be considered practical, cost-effective alternatives for managing low-risk, hydrocarbon-contaminated ground waters that are unlikely to affect drinking water wells.

PROBLEM IDENTIFICATION

Accidental releases of fuels from underground storage tanks over the past 10 to 20 years have been responsible for the presence of hydrocarbons, mainly water-soluble aromatic compounds (benzene, toluene, ethylbenzene, and xylenes, or BTEX), in aquifers. In most states, government agencies have required the regulated industry to restore ground water at such sites to drinking water (health) standards—for example, 1 to 5 parts per billion (ppb) benzene (Marencik, 1991). Corrective actions taken include removal of free product and contaminated soil, site assessments (soil borings and monitoring wells), and determination of the extent of contamination in subsoils and ground water. For a majority of the sites, the ground water hydrocarbon (BTEX) levels are low, on the order of 100 to 1000 ppb. Higher levels are often associated with soil and ground water samples taken near the spill area.

Technologies that have been used to control migration of hydrocarbon plumes or to remediate subsurface soils include soil venting (vadose zone) and sparging (saturated zone) and ground water extraction and treatment (pump and treat) (Mackay and Cherry, 1989; Newman and Martinson, 1992). In addition to these operations, extensive soil and ground water surveying must be done to assess the extent of contamination and the effectiveness of the cleanup method. Current estimates for site assessment and in situ or ex situ restoration of subsoils and ground water to health standard criteria indicate that these operations may be costly ($500,000 to $2 million per site) and not cost effective and that they may not achieve restoration within time periods of years or decades (Travis and Doty, 1990).

Many contaminated ground waters (e.g., at fuel service station sites) are in shallow aquifer zones, are not used directly for human consumption, and do not even affect downstream drinking water wells. Furthermore, good field evidence indicates that plumes in these ground waters reach a stable condition in which contaminants of concern (BTEX) are biodegraded at some rate by indigenous hydrocarbon-utilizing soil bacteria. This type of unassisted in situ biodegradation has been termed natural attenuation or intrinsic bioremediation.

Industry has been confronted with very large operating and cleanup costs for subsoils and ground water in the restoration of underground fuel storage tank sites to drinking water standards. Where thorough site characterization warrants its use, intrinsic bioremediation offers a way to manage non-migrating or shrinking BTEX plumes in low-risk aquifers that do not affect drinking water wells. Evidence that this natural process is occurring has been obtained from laboratory and field observations.

EVIDENCE FOR INTRINSIC BIOREMEDIATION OF AROMATIC HYDROCARBONS IN PLUMES

It is now widely recognized that the most significant factor in the time-dependent decrease of BTEX compounds in aquifers is degradation by soil microbes. Studies reported for laboratory slurry microcosms of subsoil and ground water show that microbes in many soils inherently biodegrade aromatic hydrocarbons at varying rates (5 to 50 percent per day) (Barker et al., 1987; Chiang et al., 1989; Gilham et al., 1990; Hutchins et al., 1991; Kemblowski et al., 1987; Major et al., 1988; and Thomas et al., 1990). These biodecay rates are usually first order; they occur with low levels of hydrocarbon (50 to 10,000 ppb); and they are rapid with adequate dissolved oxygen (e.g., 2 to 3 mg oxygen per milligram of hydrocarbon). Field estimates of hydrocarbon biodegradation rates calculated from fate and transport models using data from upstream and downstream monitoring wells have shown that plume BTEX compounds usually decrease at rates of 0.5 to 1.5 percent per day (Barker et al., 1987; Chiang et al., 1989; Kemblowski et al., 1987; Rifai et al., 1988; and Wilson et al., 1991). Laboratory and field data suggest that in a well-studied sandy aquifer a minimum, or threshold, level (≥1 to 2 ppm) of dissolved oxygen may be required to sustain hydrocarbon degradation (Chiang et al., 1989).

It should be emphasized that laboratory and field data have confirmed that all BTEX compounds can be biodegraded under aerobic conditions (dissolved oxygen in ground water) in aquifer subsoils in which oxygen is the terminal electron acceptor. Soil microcosm experiments or enrichments of aquifer material have shown that toluene and xylenes can be degraded by microbes under iron-reducing, denitrifying, and sulfate-reducing (anaerobic or very low dissolved oxygen) conditions when ferric ion (Fe^{3+}), nitrate ion (NO_3^-), and sulfate ion (SO_4^{2-}), respectively, serve as electron acceptors (Beller et al., 1992; Edwards et al., 1992; Hutchins, 1991; Lovley et al., 1989; and Zeyer et al., 1986). Field evidence is insufficient, however, to demonstrate that BTEX is biodegraded under anaerobic conditions in an aquifer.

LEVELS OF INTRINSIC ATTENUATION IN GROUND WATER

Evidence from site characterization, ground water monitoring, and modeling at field sites suggests that there may be two levels of intrinsic bioremediation. Figure 1 shows these aspects of a plume in which one is stabilized (A) and the other is reducing (B) in size and extent of contamination. In Figure 1A the hydrogeological features indicate that ground water velocity (also BTEX and dissolved oxygen) and recharge are slow because of low permeability of the aquifer subsoil. Dissolved oxygen is low within the plume (e.g., <1 ppm). Oxygen is detected in monitoring wells at the edges and is respon-

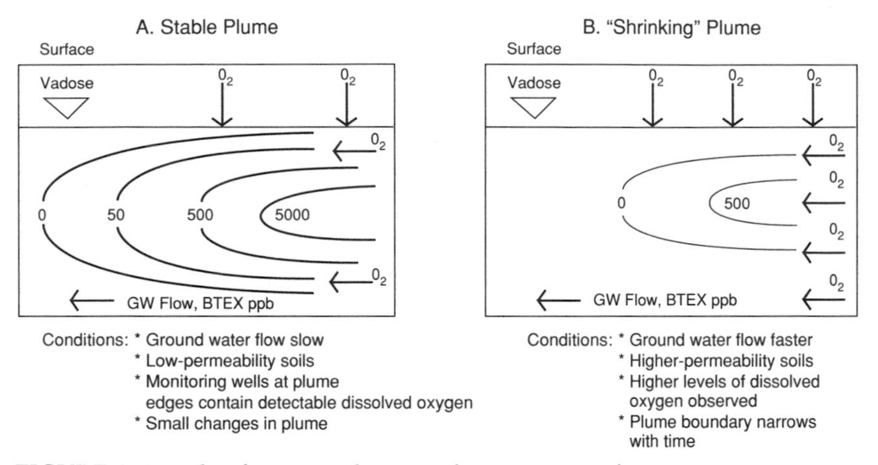

FIGURE 1 Levels of intrinsic bioremediation in aquifers.

sible for the biodegradation of low levels (ppb) of BTEX. Another indirect indicator of soil microbial degradation in aquifers low in dissolved oxygen may be the presence of dissolved ferrous ion (Fe^{2+}) above background well levels. It is known that various ferric oxides in soil can be used (as electron acceptors) by anaerobic iron-reducing bacteria to completely metabolize some aromatic compounds, such as toluene and phenol (Lovley et al., 1989). Therefore, when dissolved oxygen is low, ferric iron may substitute for oxygen, and this biodegradation process may result in elevated concentrations of ferrous ion in ground water.

At the next level of intrinsic bioremediation, plumes noticeably shrink over time, with significant decreases in shape and extent (Figure 1B). This type of plume behavior is observed in aquifers that usually are more permeable (e.g., sandy subsoils), that exhibit higher ground water velocities, and that are higher in dissolved oxygen (higher aquifer reaeration rate) in many monitoring wells. Published examples of plumes undergoing significant intrinsic attenuation of BTEX are those at the Borden (Barker et al., 1987), Traverse City (Rifai et al., 1988; Wilson et al., 1991), and Michigan gas plant (Chiang et al., 1989) sites. Monitoring wells at the periphery show significantly higher dissolved oxygen (e.g., $\geq$1 ppm) and lower BTEX concentrations, which are consistent with a predominantly biodegradation-driven mass reduction in the aquifer. Examination of monitoring well BTEX levels within the flow path of upstream and downstream segments may also match the biodecay rates (about 1 percent per day) calculated from fate and transport models for BTEX and dissolved oxygen (e.g., BIOPLUME II, Rifai et al., 1988). These plumes may initially shrink (narrow) in the longitudinal direction because the high infiltration rate of oxygen continues to enhance degradation of hydrocarbons to low concentrations at the edges. Continued monitoring also indicates that because of the higher dissolved oxygen, more BTEX is degraded and the plume may recede closer to the hydrocarbon source. It should be emphasized that the degree to which these reductions in plume BTEX occur depends on the removal of the free-phase and sorbed hydrocarbons from the contaminated zones. For example, a fluctuating water table could continue to flush more BTEX into the plume from the source area. Removal and management of the contaminant source, therefore, are important prerequisites for successfully implementing intrinsic bioremediation at field sites.

FUTURE DIRECTIONS

Laboratory research and field research have contributed to our understanding of intrinsic bioremediation of BTEX in aquifers as a viable option for managing and controlling hydrocarbon plumes. Research in several areas, however, could enhance the validity and overall regulatory acceptability of the plume containment process. For example, important factors for understanding contaminant behavior and predicting the time for remediation may include (1) a better understanding of aquifer parameters (e.g., recharge and water table fluctuations); (2) tools for quantifying subsoil sources of hydrocarbons and their potential for transport into ground water; and (3) user-friendly ground water models that use monitoring well, hydrogeological, and soil microbiology data to predict the transport and fate of contaminants. Geochemical and biological indicators of in situ biodegradation in addition to BTEX and dissolved oxygen, such as the formation of carbon dioxide and other microbial metabolites as well as ferrous ion, may also help verify intrinsic biodegradation processes in aquifers. Information on the limits of degradation of soil contaminants (e.g., optimum BTEX and dissolved oxygen concentrations and supplemental nutrient effects) and on the widespread occurrence of BTEX degraders in aquifers would also improve our understanding of plume management. Finally, it is important that demonstrated in situ biodegradation gain acceptance by the regulatory authorities and that intrinsic bioremediation be considered a valid and cost-effective means of controlling pollutant migration in low-risk aquifers. Biodegradation in aquifers will continue to play a major role in the management of low levels of soluble hydrocarbons from fuel spills to the subsurface.

REFERENCES

Barker, J. G., G. C. Patrick, and D. Major. 1987. Natural attenuation of aromatic hydrocarbons in a shallow sand aquifer. Ground Water Monitoring Review 7: 64-71.

Beller, H. R., D. Grbic-Galic, and M. Reinhard. 1992. Microbial degradation of toluene under sulfate-reducing conditions and the influence of iron on the process. Applied and Environmental Microbiology 58:786-793.

Chiang, C. Y., J. P. Salanitro, E. Y. Chai, J. D. Colthart, and C. L. Klein. 1989. Aerobic biodegradation of benzene, toluene and xylene in a sandy aquifer—data analysis and computer modeling. Groundwater 27:823-834.

Edwards, E. A., L. E. Wills, M. Reinhard, and D. Grbic-Galic. 1992. Anaerobic degradation of toluene and xylene by aquifer microorganisms under sulfate-reducing conditions. Applied and Environmental Microbiology 58:794-800.

Gilham, R. W., R. C. Starr, and D. J. Miller. 1990. A device for in situ determination of geochemical transport parameters: 2. Biochemical reactions. Ground Water 28:858-862.

Hutchins, S. R. 1991. Biodegradation of monoaromatic hydrocarbons by aquifer microorganisms using oxygen, nitrate or nitrous oxide as the terminal electron acceptor. Applied and Environmental Microbiology 57:2403-2407.

Hutchins, S. R., G. W. Sewall, D. A. Kovac, and G. A. Smith. 1991. Biodegradation of aromatic hydrocarbons by aquifer microorganisms under denitrifying conditions. Environmental Science and Technology 25:68-76.

Kemblowski, M. W., J. P. Salanitro, G. M. Deeley, and C. C. Stanley. 1987. Fate and transport of residual hydrocarbons in groundwater—a case study. Pp. 207-231 in Proceedings of the Petroleum Hydrocarbons and Organic Chemicals in Groundwater Conference. Houston: National Water Well Association and American Petroleum Institute.

Lovley, D. R., M. J. Baedecker, D. J. Lonergan, I. M. Cozzarelli, E. J. P. Phillips, and D. I. Siegel. 1989. Oxidation of aromatic contaminants coupled to microbial iron reduction. Nature 339:297-300.

Mackay, D. M., and J. A. Cherry. 1989. Groundwater contamination: pump and treat remediation. Environmental Science and Technology 23:630-636.

Major, D. W., C. I. Mayfield, and J. F. Barker. 1988. Biotransformation of benzene by denitrification in aquifer sand. Ground Water 26:8-14.

Marencik, J. 1991. State-by-state summary of cleanup standards. Soils 23:14-51.

Newman, W. A., and M. A. Martinson. 1992. Let biodegradation promote in situ soil venting. Remediation 2:277-291.

Rifai, H. S., P. B. Bedient, J. T. Wilson, K. M. Miller, and J. M. Armstrong. 1988. Biodegradation modeling at an aviation fuel spill site. American Society of Civil Engineers Journal of Environmental Engineering 114:1007-1029.

Thomas, J. M., V. R. Gordy, S. Fiorenza, and C. H. Ward. 1990. Biodegradation of BTEX in subsurface materials contaminated with gasoline. Water Science Technology 22:53-62.

Travis, C. C., and C. B. Doty. 1990. Can contaminated aquifers at Superfund sites be remediated? Environmental Science and Technology 24:1464-1466.

Wilson, B. H., J. T. Wilson, D. H. Kampbell, B. E. Bledsoe, and J. M. Armstrong. 1991. Biotransformation of monoaromatic and chlorinated hydrocarbons at an aviation gasoline spill site. Geomicrobiology Journal 8:225-240.

Zeyer, J., E. P. Kuhn, and R. P. Schwarzenbach. 1986. Rapid mineralization of toluene and 1,3-dimethylbenzene in the absence of molecular oxygen. Applied and Environmental Microbiology 52:944-947.

Bioremediation from an Ecological Perspective

James M. Tiedje
Center for Microbial Ecology
Michigan State University
East Lansing, Michigan

SUMMARY

The ecological approach to bioremediation is distinctly different from the traditional engineering approach: it focuses on such principles as microbial natural selection rather than on mass balances of pollutants. Questions derived from certain basic ecological principles, including specificity and diversity, can serve as key guides in determining the feasibility of bioremediation at a particular site. Similarly, certain kinds of evidence in the biological record, such as numbers of organisms, are strongly indicative of successful bioremediation. A shift in paradigm—emphasizing the ecological principles governing biodegradation instead of contaminant mass balances—would greatly advance the understanding of bioremediation.

INTRODUCTION

I suggest that there are at least two conceptual approaches to hazardous waste bioremediation. In the dominant approach, derived from engineering, mass balance and stirred tank reactor philosophy dominate. An alternative, or ecological, approach focuses on such principles as microbial natural selection and niche fitness characterization. Reliance on the engineering approach has brought us to an impasse—namely, that nature is not a stirred tank reactor, and thus the mass balance and predictive models of such systems are often inadequate or too expensive. In ecology, however, one recognizes from the beginning that nature is heterogeneous; to understand nature, one focuses on key principles governing the behavior of populations and does not attempt to achieve mass balances. Thus, I suggest that we consider a shift in paradigm—to consider the important ecological principles governing biodegradation and reduce the emphasis on achieving a mass balance for the pollutant.

This paper emphasizes the ecological approach and key questions related to it. The differences in the philosophies underlying the ecological and engineering approaches are substantial. As details of both approaches are developed, some of the underlying factors may merge into the same issues. Nonetheless, the emphasis in the ecological approach is not on quantification of pollutants but on whether principles are met, since it is known that biological communities respond according to these principles.

The first part of this paper reviews basic ecological principles important to the evaluation and success of in situ bioremediation. The second part converts these principles into key questions about the feasibility of bioremediation for a particular site. Finally, the paper outlines ways to determine whether the ecological principles, especially the principle of natural selection, are met.

BASIC PRINCIPLES

Specificity

An ecological approach recognizes a key principle in biology—specificity. Specificity provides the fitness advantage in a niche. In terms of pollutant degradation, this means that organisms are relatively specific for particular substrates (chemicals) and for particular environmental conditions (the niche). Oxidation by biological organisms is the extreme opposite of oxidation by combustion. The former is specific for particular chemicals, while the latter is entirely nonspecific. The specificity of biological organisms is conferred by such features as membrane selectivity, permeases, regulatory proteins controlling enzyme synthesis, and the structure of the enzyme-active site. There is too great a tendency to generalize about bioremediation as a class of technology, like combustion, which obscures the fact that biodegradability should *always* be discussed together with the particular chemical.

Although specificity may seem to be a disadvantage for bioremediation, in fact it provides one of the cost advantages of the technique because resources are focused *only* on the target chemical. In combustion, for example, external resources are needed to oxidize all organic compounds, while in biodegradation resources go only to the compounds that can reach the enzyme's catalytic site. In cometabolism, where external resources are often needed, this feature is extremely important.

Microbial Diversity

Diversity, nature's counter to specificity, results from evolution, in which organisms diversify from their progenitors to occupy new niches. Because of the heterogeneity in nature, there are many niches and thus a naturally high degree of biodiversity. For bacteria, diversity seems to be exceedingly high; there are likely more than 10,000 species per gram of soil (Torsvik et al., 1990). Fungi also seem to be very diverse, with an estimated 1.6 million species on earth (Hawksworth, 1991). Most of these organisms have never been isolated, let alone studied. For example, *Bergey's Manual*, which describes all known bacteria, includes only 3000 to 4000 species, and most of these are not from soil or water (Holt, 1989).

This great diversity is important to bioremediation in two ways. First, it means some diversity in the mechanisms that confer specificity. For example, a small number of the organisms that degrade benzoate will also be able to degrade chlorobenzoate or perhaps dichlorobenzoate, because the active site pocket is slightly modified in these variants to allow access to the bulkier chlorine group. This principle seems to be important in the metabolism of polychlorinated biphenyls (PCBs), since the oxygenase of some toluene- and naphthalene-degrading organisms can attack PCBs (Kuhm et al., 1991). Generally, the principle applies to structurally similar chemicals or chemicals subject to the same mechanism of attack. Thus, specificity is not absolute but usually limits the range of substrates attacked to very few.

The second reason that diversity is important is that it is thought to lead to a more robust and stable process because diverse species are likely to include specialists for assimilating low and high pollutant concentrations; for tolerances for different pHs, metals, and solvents; for different growth rates; and for different resistances to phage infection or protozoan grazing. For example, among benzene degraders in a gram of soil, there may be hundreds or even thousands of indigenous strains that may vary in these other important ecological traits. Original ecological dogma was that more diversity leads to stability, but current evidence from macroecology suggests that less complex systems are more stable (e.g., Begon et al., 1990). However, no evidence exists on the relationship between stability and diversity for a microbial process. In any case, higher diversity among pollutant degraders should lead to emergence of the most fit organisms for the degradation and hence enhance degradation performance.

If high diversity and large populations of pollutant degraders already exist in the habitat, it becomes virtually impossible to successfully introduce an inoculum. The native organisms both preemptively colonize the niche and are likely more fit for the niche. Thus, super biodegraders, whether natural or genetically engineered, stand little chance against a significant indigenous population that can degrade the target chemical.

Biogeography of Biodegraders

Bacteria have been on earth for 3 billion years, an extremely long period of time. Indeed, 85 percent of bacterial existence to date occurred before the continental plates began to drift apart. Thus, the organisms have had a very long time to evolve, adapt, and disperse. This long period likely also led to excellent survival strategies, so that organisms can persist outside their optimum niches for many years. A century ago, Beijerinck, a famous Dutch microbiologist, stated that "everything [bacterial types] is everywhere, the environment selects." This remains the accepted dogma. Extended to biodegrading organisms, this dogma suggests that biodegradative traits found in one soil or water would be found in most other soils or waters around the world. The global distribution of such traits has not yet been fully evaluated (and is the subject of research), but general experience suggests that the dogma is true, at least at the level of the particular activity, if not the identical strain. Hence, there may be some local variation, but it likely occurs at the variety or strain level and is probably not apparent at the process level. In other words, biodegradation proceeds on similar substrates and at similar rates even though some of the strains are slightly different.

The importance of this biogeographical analysis to bioremediation is that it suggests that biodegrading populations are similar at many sites. The portion of biodegrading organisms in the total community at a given site may be somewhat similar to that at other sites if selection has not already occurred. Thus, if the total population is high, as in a fertile surface soil, the biodegrading population will be high. In contrast, in the vadose zone and aquifer soil, which are impoverished in organic matter, the total populations will be lower and hence so will be the biodegrading populations. At sites where biodegrading populations are likely to be large based on similarity to other sites, it would be difficult to successfully inoculate a biodegrading organism.

Pollutants as Analogs of Natural Products

Biodegradation occurs when organisms have enzymes that can attack the substrate. Natural selection throughout evolutionary history has maintained those enzymes because they enhance fitness. Thus, pollutant degradation occurs because this enzyme probably also metabolizes an analogous natural product in order for selection to have preserved the gene sequence. It is often very difficult to identify the natural substrate for the biodegrading enzyme without obvious structural analogs. For example, halogenated chemicals are rare in nature, and the natural substrates for enzymes involved in reductive dehalogenation are completely unknown (Mohn and Tiedje, 1992).

The corollary of this situation is that bond types (or structures) not known in nature are often not metabolized. Since these new substrates are a potential energy resource, they exert selective pressure for organism variants to use them. To acquire basically new enzymatic traits through natural evolution is thought to take a very long time, probably hundreds or thousands of years. If one wants to biodegrade these nonnatural chemicals in our lifetime to clean up hazardous waste, the task will likely involve protein and gene engineering, a process not financially feasible in the foreseeable future.

Natural Selection

Ecological systems are driven by the resources available and the competition for them among the community members. For pollutant degradation, the major question is whether the pollutant is an energy resource—will an organism grow on the chemical as a substrate? If so, there is strong selective pressure for the degrading population to outgrow others, thereby amplifying the rate of degradation. It is useful to group chemicals into two classes of biodegradability: (1) those that support the growth of microbial populations and (2) those that are cometabolized (in other words, they do not support growth but are partly metabolized, usually through only a step or two of the complete metabolic pathway). Organisms that carry out cometabolism are not naturally selected and, therefore, are much more difficult to manage in nature. For this reason the distinction of these two classes is important.

When pollutants are growth substrates, major advantages accrue: (1) the catalyst grows logarithmically with no external input of resources; (2) the proper growth, activity, and distribution of the microbial population (which is very difficult to manage under other circumstances) is an inherent outcome of natural selection for the primary energy substrate; and (3) growth substrates are almost always completely oxidized to carbon dioxide, leaving no toxic intermediates. Less than complete pollutant destruction by natural selection is usually due to limitation by some other resource, most commonly the electron acceptor. Because of these advantages, chemicals that are growth substrates have not and should not become widespread pollution problems. This is because the limitations on natural selection disappear as the chemical becomes more widely distributed. Examples of chemicals that are growth substrates are benzene, toluene, xylenes, naphthalene, chlorophenols, acetone, nitrilotriacetic acid, and 2,4-D. Whenever a pollutant is a growth substrate, bioremediation should be seriously considered. Even if the waste contains mixtures of chemicals, some of which are growth substrates and others not, bioremediation may still be advantageous because it can reduce other remediation costs, such as the amount of activated carbon needed.

Cometabolism usually results from relaxed specificity of an enzyme. No sequential metabolic pathway or energy coupling to adenosine triphosphate production typically occurs. Therefore, natural selection cannot be achieved through this secondary (pollutant) substrate. If cometabolism is to be used, it must be done by managing a primary substrate that selects for growth of active organisms, induces the activity, and/or provides a necessary oxidant or reductant to drive the reaction. Sometimes the primary and secondary substrates are competitive inhibitors, which may require more sophisticated management, such as pulse feeding or precise concentration control. Cometabolic processes typically accumulate intermediates, some of which may be toxic.

Cometabolic reactions seem to be the only ones that show activity on many of the recalcitrant chlorinated solvents, such as perchloroethylene (PCE), trichloroethylene (TCE), carbon tetrachloride, and chloroform. Laboratory testing and field testing are beginning to show that it may be possible to successfully manage a cometabolic process in situ. Nonetheless, the experimentation, field testing, and monitoring will all need to be more extensive than for pollutants that are growth substrates.

THREE KEY QUESTIONS

I suggest that the following key questions, in the indicated order of priority, are a basic guide to successful bioremediation:

1. Is the chemical degradable?
2. Is the environment habitable?
3. What is the rate-limiting factor and can it be modified?

Is the Chemical Biodegradable?

The first question is whether the chemical is biodegradable, because bioremediation cannot be accomplished if no organism exists that can degrade the chemical. Biodegradability must be established if it is not already well documented in the literature. Subquestions are whether the chemical is a growth substrate, for the reasons discussed above, and whether the biodegrading organism exists at the site.

A focus on the biodegradability of the pollutant is also important because it suggests the time until application and the research needed for application, as shown in Figure 1. In the figure, biodegradability is indicated by the frequency of the biodegrading populations within the total soil community. The higher frequency implies several benefits to bioremediation, including greater diversity among the populations of degraders, less chance of encountering patches devoid of organisms, and a rather global distribution of this biodegradative property at most sites, which allows extrapolation of information among sites. If organisms are widespread, they cannot be limiting to biodegradation. Hence, environmental factors are then the focus for ensuring or enhancing bioremediation.

The time until field application of a bioremediation technology can also be predicted by the biodegradability scale of Figure 1. When natural degrading organisms are widespread, application is more immediate because conditions may be met naturally or, if not, technology exists for removing some of the environmental limitations. However, when organisms do not exist or are rare, the time until application is more distant because successful addition or distribution of organisms is difficult to achieve, especially in the subsurface (Harvey et al., 1989). It is even more difficult to genetically engineer a new catalytic property; this approach is far from any practical application to bioremediation.

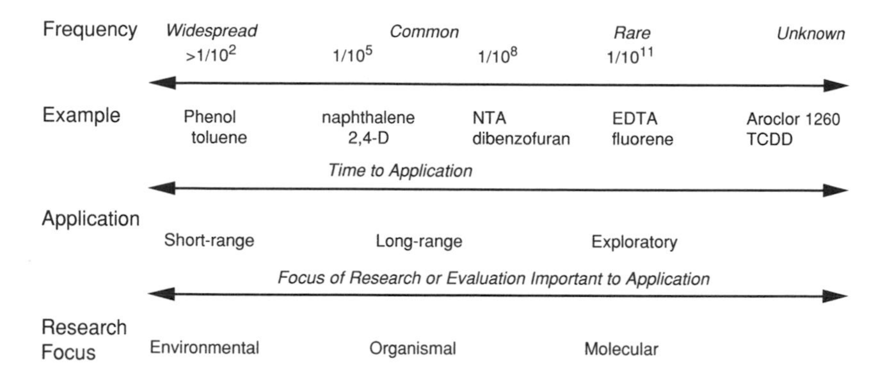

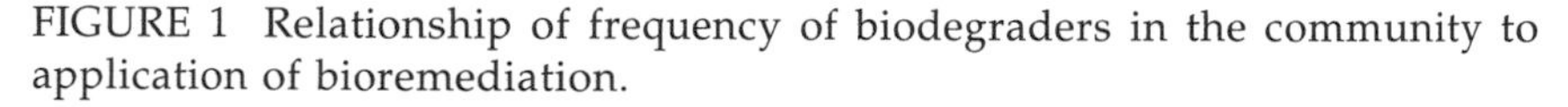

FIGURE 1 Relationship of frequency of biodegraders in the community to application of bioremediation.

Is the Environment Habitable?

The second question—is the environment habitable?—comprises two issues. First, does the environment contain toxic chemicals that make it difficult or impossible for microbes to live? Many polluted sites contain mixtures of chemicals and metals, some at high concentrations, that may pickle the environment so that bioremediation is not feasible. The second issue is the availability of sufficient life-sustaining growth factors, such as nutrients, particularly nitrogen and phosphorus; appropriate electron acceptors; and perhaps other growth factors that might be contained in soil organic matter. Nutrient supply can be evaluated by considering whether the proper carbon-nitrogen-phosphorus (C:N:P) ratio is likely to be met by the soil environment. A C:N:P ratio of 30:5:1 is needed for unrestricted growth of soil bacteria (Paul and Clark, 1989). Microbial growth in most subsoils is not limited by nitrogen and phosphorus as long as the new carbon being provided is not in amounts greater than tens of parts per million. This is often the case with pollutant chemicals. Since nitrogen and phosphorus are inexpensive, however, they are often added as insurance.

What Is the Rate-Limiting Factor and Can It Be Modified?

Too often in bioremediation there is a solution in need of a problem. Thus, effort or money is spent to modify something that is not rate limiting. To avoid this waste, the rate-limiting parameter must first be defined. In doing so, it is worthwhile to consider the ecosystem in its entirety and to recognize the three key components: sub-

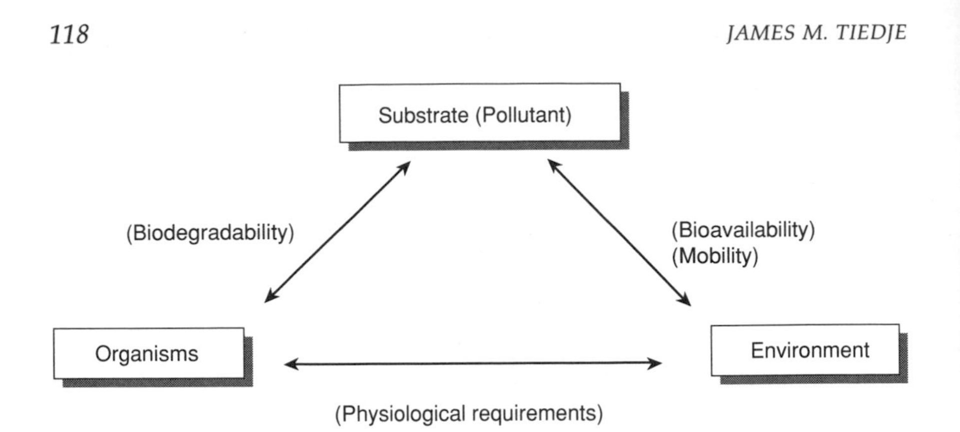

FIGURE 2 Interrelationships of essential components that determine successful bioremediation.

strate (pollutant), biodegrading organism, and environment, as depicted in Figure 2. Factors that reflect this interrelationship and that can limit biodegradation are shown in parentheses in Figure 2.

If biodegradability and habitability have been established, the most common limiting parameter is oxygen, since it has relatively low solubility in water and is in high demand as an oxidant for all biological respiration. Thus, schemes for injection of oxygen or hydrogen peroxide into soil or aquifers are common. Such treatments overcome a rate limitation if the site is anaerobic. Alternative electron acceptors are possible, and nitrate is particularly attractive because of its high electron-accepting capacity in water, its leachability in soils, its low toxicity, and its low cost. Research on denitrification-driven bioremediation is in its infancy, however. The frequency of this type of biodegrading population in soil is not known, but it almost certainly is lower than for oxygen-respiring organisms.

Other treatments to meet physiological requirements include addition of nutrients, adjustment of pH, and removal of toxicants by leaching, precipitation, or some form of inactivation. As stated above, nutrient addition is common, probably because it is easy and cheap and may occasionally provide some benefit, not because it has been a well-documented requirement for many sites.

A second important limitation on biodegradation is the availability of the chemical to the organisms, or bioavailability. Bioavailability is limited when the pollutant is dissolved in organic matter or trapped in micropores in the soil matrix. Substantial work is under way to attempt to understand and enhance the bioavailability of water-insoluble chemicals. The ecological approach to this problem, however, would be to focus on ensuring that the local environment contains zones that would support natural selection if and when the chemical became available, and not on the immediate (and impossible) recall of that chemical from all microsites.

A related issue, but on a slightly larger scale, is the movement of the chemical or organism so that the two come into contact. Mobility is not a limitation for water-soluble chemicals, which move through soil easily, but it is a severe problem for very insoluble chemicals. In this case, movement of organisms is all that is feasible if physical mixing is not possible.

CONCLUSION

Returning to the ecological approach, the key point in determining whether bioremediation is successful is to establish *whether the conditions of natural selection can be expected to be met within the site vicinity*. The point is not to determine pollutant mass balances; it is not to ensure that all heterogeneity can be understood and accounted for; and it is not even to worry about local concentrations above regulatory targets if conditions of the surrounding environment ensure that natural selection will occur. This approach recognizes that energy from organic matter is the key limitation for microbial growth and that if the appropriate enzymes and required environmental conditions exist, there is no way to prevent complete biodegradation. Thus, the first criterion for successful bioremediation is documentation of the conditions for natural selection, namely: (1) is the chemical a growth substrate for microbes? (2) is the site habitable (nontoxic) for microbial life? (3) is there sufficient electron acceptor? The ecological approach suggests that more emphasis should be placed on documenting adequate electron acceptor supply and less on measuring the actual pollutant.

A second line of evidence for a successful bioremediation is whether the biological record suggests that natural selection has occurred. This evidence was well illustrated by Madsen et al. (1990) for a plume from a coal tar site. Types of evidence in the biological record include (1) increased rate of pollutant mineralization; (2) increased populations of microorganisms (e.g., total microbial populations, the biodegrading population, and grazers of those populations); and (3) chemical gradients that show a discontinuity caused by respiratory consumption of electron donors (pollutant) and electron acceptors. At contaminated sites, this kind of evidence in the biological record would be strongly indicative of successful intrinsic bioremediation and its persistence as long as the conditions for natural selection can be ensured.

ACKNOWLEDGMENTS

The author's research on biodegradation has been funded by the U.S. Environmental Protection Agency and the National Institute of Environmental Health Sciences Superfund Program.

REFERENCES

Begon, M., J. L. Harper, and C. R. Townsend. 1990. Ecology: Individuals, Populations and Communities. Cambridge, Mass.: Blackwell Scientific Publications.

Harvey, R. W., L. H. George, R. L. Smith, and D. R. LeBlanc. 1989. Transport of microspheres and indigenous bacteria through a sandy aquifer: results of natural- and forced-gradient tracer experiments. Environmental Science and Technology 23:51-56.

Hawksworth, D. L. 1991. The fungal dimension of biodiversity: magnitude, significance, and conservation. Mycological Research 95:641-655.

Holt, J. G. 1989. Bergey's Manual of Systematic Bacteriology. Baltimore: Williams & Wilkins.

Kuhm, A. E., A. Stolz, and H. J. Knackmuss. 1991. Metabolism of naphthalene by the biphenyl-degrading bacterium *Pseudomonas paucimobilis* Q1. Biodegradation 2:115-120.

Madsen, E. L., J. L. Sinclair, and W. C. Ghiorse. 1990. In situ biodegradation: microbiological patterns in a contaminated aquifer. Science 252:830-833.

Mohn, W. W., and J. M. Tiedje. 1992. Microbial reductive dehalogenation. Microbiological Reviews 56:482-507.

Paul, E. A., and F. G. Clark. 1989. Soil microbiology and biochemistry. San Diego: Academic Press.

Torsvik, V., J. Goksoyr, and F. L. Daae. 1990. High diversity in DNA of soil bacteria. Applied and Environmental Microbiology 56:782-787.

In Situ Bioremediation: The State of the Practice

Richard A. Brown
Groundwater Technology, Inc.
Trenton, New Jersey

William Mahaffey
ECOVA Corporation
Redmond, Washington

Robert D. Norris
Eckenfelder, Inc.
Nashville, Tennessee

SUMMARY

Since the pioneering work by Dick Raymond during the 1970s and early 1980s, in situ bioremediation has been widely used to clean up aquifers contaminated with petroleum hydrocarbons. A need for better performance led to development of the use of hydrogen peroxide and direct injection of air into the aquifer as sources of oxygen, which was a critical problem in bioremediation. Bioremediation has developed in two branches. The first has been engineering techniques and mathematical models for applying bioremediation to readily degradable contaminants. The second branch has focused on ways to address more recalcitrant contaminants such as chlorinated solvents, polychorinated biphenyls, and pesticides. Work on these more challenging problems has met with some success in the laboratory, but the techniques have yet to be commercialized, largely because of failure to establish and maintain critical control parameters in the subsurface. Continued improvements in the technology will result from efforts in site delineation, engineering controls, use of nonindigenous microorganisms, and field methods for evaluating the microbiological processes.

INTRODUCTION

Bioremediation was first used commercially in 1972 to treat a Sun Oil gasoline pipeline spill in Ambler, Pennsylvania (Raymond et al., 1977), and has been used almost as long as simple pump-and-treat technology. In situ bioremediation was one of the first technologies that was able to bring a site to closure by significantly and permanently reducing soil and ground water contamination, predating in situ processes such as soil vapor extraction and air sparging.

The evolution of in situ bioremediation has had three important aspects: microbiology, engineering, and applications. The microbiological aspects have been concerned with basic metabolic processes and how to manipulate them. Much of this work has been and continues to be laboratory scale and is currently directed at recalcitrant substrates such as polychlorinated biphenyls (PCBs), chlorinated solvents, and pesticides. The second aspect, the engineering of in situ bioremediation, has been concerned with field-scale systems needed to provide the substances required for the metabolic processes, such as oxygen, moisture, and nutrients (Brown and Crosbie, 1989). The most difficult aspect of development has been the translation of laboratory results to field applications. Finally, specific types of bioremediation have been developed to treat specific types of contaminants or matrixes. For example, a significant outgrowth of in situ bioremediation has been the development of ex situ soil biotreatment (Brown and Cartwright, 1990), which has become a cost-effective and widely applied on-site technology. The engineering aspects of bioremediation have produced the greatest successes in the commercial use of the method, leading to the development of specific applications.

Bioremediation has been a successful technology when properly used. It is also an oversold technology, having more promise than results. Understanding the practice of in situ bioremediation—its legitimate uses and potential results—requires an examination of historical developments in microbiology, the current status of the practice of bioremediation, and new developments in bioremediation. This examination illustrates the successes, limitations, and continued needs of bioremediation technology.

HISTORICAL DEVELOPMENTS

The development of bioremediation has been predicated on an evolving use of indigenous microorganisms to biodegrade a variety of organic compounds in soils and wastewater. A large body of information about biooxidation mechanisms and products and the effects of reaction conditions was available before the technology was commercialized. The microorganisms that could degrade various classes of compounds under both aerobic and anaerobic conditions and the effects of and requirements for pH, nutrients, oxygen, temperature, redox potential, and moisture were all reasonably well established before in situ bioremediation was practiced commercially.

Early studies in hydrocarbon metabolism were reported by Tausson (1927), who isolated bacterial strains capable of oxidizing naphthalene, anthracene, and phenanthrene. Subsequently, Sisler and Zobell (1947) demonstrated that marine bacteria could rapidly oxidize benzo[*a*]anthracene to carbon dioxide. Senez and co-workers (1956) were the first to suggest that normal alkanes were enzymatically attacked at the first carbon atom (C1 position). Finally, Leadbetter and Foster (1959) were the first to observe, define, and report on the cooxidation of hydrocarbons previously considered resistant to oxidation and assimilation.

Early in the development of bioremediation, oxygen availability was seen as a critical factor (Floodgate, 1973; Zobell, 1973). The concept of introducing water amended with nutrients and oxygen (using in-well aeration) to promote biodegradation was first tried by Dick Raymond in 1972 at the Ambler pipeline spill mentioned earlier. This technology was patented by Raymond in 1974.

From 1975 to 1983, Raymond and co-workers (Jamison et al., 1975) conducted several demonstration projects with the support of the American Petroleum Institute (API). These studies demonstrated the feasibility of in situ bioremediation; the observed reductions in soil and ground water contamination were sufficiently encouraging to stimulate widespread interest in the technology. This early work identified oxygen supply as crucial if the technology was to be generally applicable. This finding led to the innovative use of hydrogen peroxide as an oxygen carrier (Brown et al., 1984).

Laboratory tests at the Texas Research Institute (1982) demonstrated that hydrogen peroxide could be a source of oxygen for bacteria and could be tolerated at concentrations up to 1000 mg/l. API and FMC Corporation supported a field test in Granger, Indiana, that demonstrated that hydrogen peroxide could be used on a field scale (American Petroleum Institute, 1987). The use of hydrogen peroxide as an oxygen source and as an agent for maintaining well performance was subsequently patented (Brown et al., 1986).

During 1983-1986, several commercial in situ bioremediation projects using hydrogen peroxide as the oxygen source were implemented and in some cases reduced hydrocarbons (Frankenberger et al., 1989) and BTEX (benzene, toluene, ethylbenzene, xylenes) (Norris and Dowd,

1993) to below detection limits. Because of the potential for more efficient oxygen supply, the use of hydrogen peroxide expanded interest in bioremediation. However, even though hydrogen peroxide did significantly improve oxygen supply, it, too, had severe limitations: in the treatment of vadose zone (unsaturated) soils and the instability of hydrogen peroxide in certain types of soils (Britton, 1985), which can cause problems such as too rapid decomposition and formation plugging.

The first change in the use of hydrogen peroxide came with the development of soil vapor extraction (SVE), which is now recognized as a more efficient supplier of oxygen for unsaturated soils and which has replaced the use of hydrogen peroxide (Brown and Crosbie, 1989). While the focus of soil vapor extraction has always been removal of volatiles, it was observed that the process of vapor recovery could also result in substantially increased biodegradation rates (Thornton and Wooten, 1982; Wilson and Ward, 1986). Several recent tests, such as those conducted by the U.S. Air Force, have demonstrated a high degree of biooxidation versus physical removal (Miller et al., 1990).

The development of SVE led to a broadening of remedial technology. Because soil vapor extraction could physically remove volatile organics, bioremediation became less of a stand-alone technology. Site remediation became an integrated approach using SVE and bioremediation.

Concerns with hydrogen peroxide stability led to a search for other soluble electron acceptors. Several tests were conducted to evaluate nitrate as an alternate electron acceptor for degradation of monoaromatic (except benzene) and polyaromatic compounds. Nitrate is inexpensive, is easily transported through the formation, and appears to cause fewer problems than oxygen. However, nitrate does not result in degradation of aliphatic compounds, and its use may be limited by state and local regulations and concerns for nitrite formation and potential for eutrophication.

The most recent innovation in bioremediation technology has been the use of air sparging to oxygenate ground water (Brown and Jasiulewicz, 1992). Air sparging involves injecting air below the water table to saturate the ground water with air (and thus provide oxygen), as shown in Figure 1. The process can also transfer volatile components to the unsaturated zone for capture by a vapor recovery system. Currently, air sparging is receiving great attention because it is relatively inexpensive and can distribute oxygen across the entire site at one time rather than relying on an oxygen front moving across the site. In formations where air sparging is applicable, it has supplanted hydro-

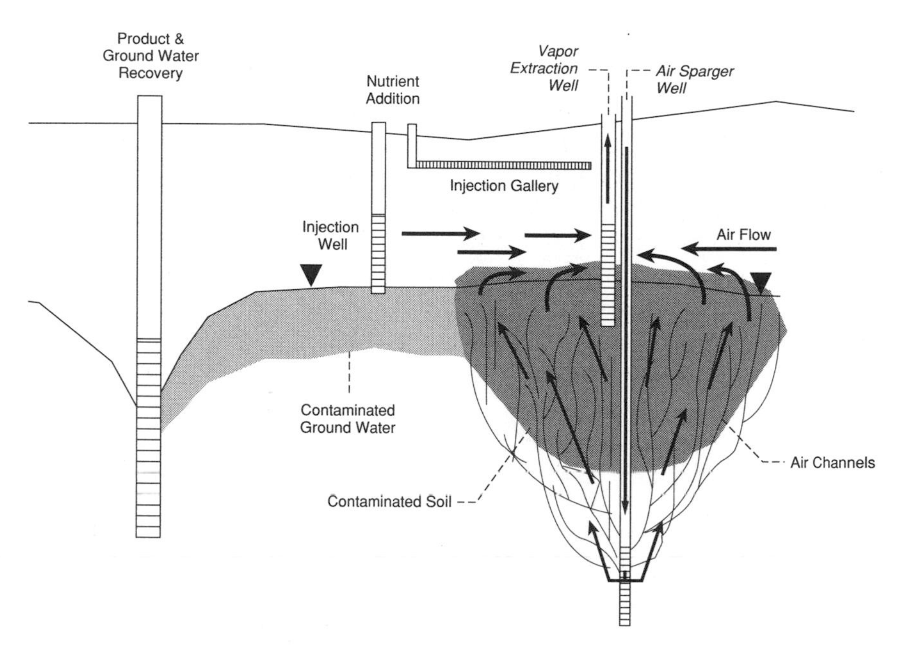

FIGURE 1 Diagram of integrated remedial system.

gen peroxide. Air sparging provides the same benefits to saturated zone treatment that soil vapor extraction has to vadose zone treatment.

CURRENT USES

The application of bioremediation is continually changing. Initially, the technique was viewed as a primary treatment process—able, potentially, to treat a wide range of organic compounds in soil and ground water. The advent of soil vapor extraction and air sparging, however, has diminished the importance of bioremediation as a standalone system for contaminants that are relatively volatile and thus readily removed physically by sparging and venting. As a result, bioremediation has evolved in two directions: as part of an integrated system for treating highly mobile (volatile and/or soluble) and/or degradable substrates, such as gasoline or diesel fuel, and as a primary system for treating nonmobile or recalcitrant substrates such as heavier petroleum products and, potentially, PCBs and pesticides.

Treatment of Degradable Mobile Contaminants

In the integrated treatment of hydrocarbon fuels or other mobile and degradable substances, bioremediation, or biodegradation, has become an effective incremental technology in conjunction with SVE and air sparging. Biodegradation occurs readily during the aeration of petroleum hydrocarbons (Miller et al., 1990). The degree to which biodegradation occurs relative to other removal processes, such as volatilization, depends on the properties of the contaminant and the rate of air flow and other environmental factors. Biodegradation can be enhanced by adjusting air flow and moisture and by adding nutrients. Physical removal is enhanced by increasing air flow.

The design of bioremediation strategies is highly site specific. It depends on contaminant properties and distribution, lithology, infrastructure (buildings, pavement, utilities, etc.), regulatory requirements, and client-specific issues such as site usage and time requirements. For instance, soil permeability and layering of highly permeable or very tight soils may preclude one or more technologies or restrict design options. Generally, most in situ processes have had little success in clay-based soils.

Many sites are now being remediated using multiple technologies. Where free-phase hydrocarbons are present, it is almost always advisable to remove the recoverable free-phase liquids. This typically leaves small pockets of free-phase liquids as well as soils contaminated with several thousand parts per million of adsorbed-phase organics. Pump-and-treat methods will satisfactorily remove only those contaminants with water solubilities in excess of 10,000 mg/l. Thus, remediation of most sites requires the incorporation of technologies that can remove or destroy substantial quantities of contaminants.

For volatile biodegradable contaminants, a combination of in situ bioremediation, air sparging, and/or vapor extraction may be the best strategy, provided the soil properties and site infrastructure permit. Designs that emphasize air sparging and vapor recovery are likely to lead to faster remediation than systems that emphasize bioremediation. The latter, accomplished by using intermittent or low air flow rates, offers the advantage of minimizing off-gas treatment as a trade-off for speed of remediation.

Integration of technologies will typically provide the most cost-effective remedial design. Thus, where unsaturated soils are contaminated by biodegradable substances with vapor pressures exceeding approximately 1.0 mm Hg, a combination of vapor recovery and bioremediation is likely to be used. Where saturated zones are contaminated largely by compounds with vapor pressures exceeding 1.0 mm Hg and with Henry's Law constants exceeding 10^{-5} atm·m^3/mole, air sparging can be used to provide oxygen and physically transfer contaminants to the unsaturated zone for capture with a vapor extraction system.

For biodegradable contaminants with minimal volatility, bioremediation may be a stand-alone technology. Polyaromatic hydrocarbons (PAHs), heavy fuels, and plasticizers, for example, respond primarily to bioremediation alone. The oxygen, however, may be provided by air sparging and/or vapor extraction techniques. In fractured bedrock, highly stratified aquifers, or where the saturated interval is no more than about 1 m, oxygen is more aptly provided through recirculated ground water using hydrogen peroxide.

Resistant Organics

Recent years have seen continued progress with microbial degradation of chlorinated solvents, pesticides, PCBs, and nitroaromatic compounds. In general, however, the current state of technology does not permit these classes of compounds to be treated on a commercial scale. Similarly, there is little evidence that nonindigenous microorganisms have been used successfully on a commercial scale for in situ bioremediation.

With highly degradable substances, intrinsic bioremediation can be used as the final treatment when the contaminant load has been reduced to the point that the ambient nutrient levels and oxygen diffusion are sufficient to support biodegradation. With this unassisted bioremediation, treatment costs can be very low.

FUTURE OF THE TECHNOLOGY

The engineering aspects of bioremediation have produced the greatest successes in commercial use of the methods. This is due primarily to a substantial body of information that existed on microbial use of petroleum hydrocarbons as sources of carbon and energy for growth. Raymond's pioneering efforts in the commercialization of bioremediation for petroleum hydrocarbons were based on 45 years of research in biodegradation. In considering the future of bioremediation it is wise to maintain perspective on the historical elements of microbiology and biotechnology that support the engineering breakthroughs. One must also acknowledge the current technical limitations of in situ bioremediation, which fall into three major and highly interactive areas: physical/chemical, microbiological, and site assessment.

Physical/Chemical Limitations

Major engineering advances have already been made in overcoming physical/chemical constraints on in situ bioremediation systems, particularly in the area of oxygenation. However, certain physical/chemical elements still significantly affect the microbiological component of in situ bioremediation. Of these, the molecular architecture of organic pollutant molecules has the greatest implications.

Size and the extent and type of functional group substitution dictate the bioavailability and biodegradability of a molecule. Bioavailability through desorption is greatly reduced by solubility limitations as well as degree of hydrophobicity, both of which depend on molecular size and functional group substituents. Surfactants may improve bioavailability, but they are of no avail where the microbial populations lack the catabolic capacity to biodegrade the molecule(s) of concern.

Another important factor is that single-substance contamination is rare in most polluted environments. Microbial biodegradation of multicomponent mixtures is not as well understood as many would believe. Biodegradation of complex mixtures is often assumed to occur if the contaminants are known to be biodegradable and substrate interactions are known to be not important. However, at least two studies involving gasoline (Barker et al., 1987; Wilson et al., 1990) reported that some BTX (benzene, toluene, xylenes) constituents persisted above regulatory action levels, even after stimulation of bioremediation by addition of inorganic nutrients and various electron acceptors. A number of investigators (Alvarez and Vogel, 1991; Arvin et al., 1989; Bouwer and Capone, 1988) have recognized and begun to investigate the importance of substrate interactions.

Microbiological Limitations

The unpredictability of biodegradation adds to the importance of continued research on metabolic processes such as adaptation, cooxidation, diauxy, catabolite repression, and competitive inhibition. Central requirements of in situ bioremediation are that the contaminants are biodegradable, that the appropriate microbial populations are present, and that the microbes are able to thrive. The understanding of metabolic pathways in biodegradation and of the factors that control microbial populations continues to grow, thus increasing the potential for bioremediation.

Research into the biodegradation of chlorinated organics illustrates the importance of continued microbiological research. Chlorinated solvents and many other halogenated compounds (e.g., PCBs) were previously thought to be recalcitrant both aerobically and anaerobically. However, the early 1980s witnessed major advances in our fundamental knowledge of the biodegradation of chlorinated organics. Bouwer et al. (1981) demonstrated the anaerobic degradation of halogenated 1- and 2-carbon compounds. Subsequent research on trichloroethylene (TCE) (Vogel and McCarty, 1985) and perchloroethylene (Fathepure et al., 1987) demonstrated that these compounds were cometabolized through a reductive dehalogenation mechanism by a consortium of anaerobic organisms. Researchers at General Electric Corporation (Bedard et al., 1987; Quensen et al., 1988) identified a reductive dehalogenation mechanism for PCBs. Bedard and her coworkers further demonstrated novel aerobic processes that degraded the more refractory orthosubstituted PCB congeners and have isolated a number of bacterial strains that are highly efficient in degrading the more highly chlorinated congeners. TCE was shown to be cooxidized by methanotrophic bacteria supplied with methane (Wilson and Wilson, 1985) and by a strain of *Pseudomonas cepacia* (G4) supplied with phenol or toluene (Nelson et al., 1987).

There has been a plethora of laboratory investigations to identify beneficial microbial processes but relatively few field pilot studies demonstrating the efficacy of in situ bioremediation for recalcitrant compounds and little commercialization of novel microbial processes. Extensive field studies by researchers at Stanford University used stimulation of methanotrophs to cooxidize TCE under nearly ideal field conditions. To date, the technology has not been commercialized. Another in situ field study was performed by General Electric Corporation in the summer of 1991. While limited in scope, this study provided field-scale data for evaluating aerobic biodegradation of PCBs by naturally occurring microorganisms. Despite these partially successful field studies, there has been little progress toward commercialization of new bioremediation processes for in situ application.

The disparity between research success and commercialization reflects the difficulty of maintaining critical control parameters (e.g., the requirement of methane for cooxidation of TCE and the coincident competitive inhibition of TCE degradation in the presence of excess methane). Further, many research studies use highly adapted cultures that are not readily dispersed throughout the formation or maintained in the presence of predators. To date, there has been only a preliminary report suggesting that the injection of a specific degrader population, *P. cepacia* strain G4, for cooxidizing TCE may be effective under highly ideal site conditions (Nelson et al., 1990). These results are currently being reevaluated by further field pilot testing.

Site Assessment Limitations

An important element of any field pilot program is that the site be well characterized and that statistically valid sampling plans be used during the site investigation and remediation. Several critical elements of an environmental sampling plan are:

- a definition of the time-space population(s) of interest;
- development of field-sampling designs and sample measurement procedures that will yield representative data from the defined populations; and
- assessment of the uncertainty of estimated quantities through means, trends, and average values.

Evaluation of the applicability of bioremediation requires answers to some basic questions, such as: What is the validity of assuming that the enumeration of specific degrader populations can be used to assess the degradative potential at a site or that these populations can be adequately stimulated to degrade the pollutants? How many site samples must be analyzed by treatability methods to demonstrate a biodegradative rate enhancement sufficient to achieve a regulatory level? What are the acceptable standardized methods? How well do the data from these test methods predict actual field results, and do these results justify the costs of obtaining these data? The answers to these questions are likely to vary from site to site and will be greatly influenced by experimental design. Field performance can be predicted from laboratory experiments only through development of appropriate mathematical models that are verified over time by demonstrating good correlation of laboratory and field data.

Future Needs

The future of bioremediation lies in overcoming the limitations of the technologies. Clearly, the enormous costs of site remediations and the goal of eliminating future liability constrain the development of new technologies. The most significant advances will be those that result in the development of predictable, efficient, lower-cost methods of remediation. Some of the limitations are physical/chemical and will be overcome by purely engineering methods; other solutions will be uniquely biological. In addition to the identification of new microbial capabilities for degrading chemical pollutants, other biotechnical offshoots will evolve. These can be viewed as bioaugmentation, analytical methods, and process innovations.

Bioaugmentation

Genetic engineering to improve catabolic capacity has enormous potential for obviating cellular regulatory control over the expression of biodegradative pathways. This technology offers the distinct advantage of constructing new biodegradative pathways by eliminating misrouting of metabolites to end products that inhibit further biodegradation of a pollutant (Reineke and Knackmuss, 1990). The use of specially constructed strains to biodegrade a heretofore recalcitrant pollutant would expand the range of compounds and therefore the number of sites amenable to bioremedial technologies. However, until the release of genetically engineered organisms is more acceptable from a social and regulatory perspective, this technology will be of use only from an academic perspective.

An alternative to classical genetic engineering is laboratory breeding of organisms under appropriate selective pressures to enrich for strains with the desired phenotypic characteristics. This process was effective in isolating a single strain of bacteria capable of degrading chlorobenzenes from the coculture and in the selective breeding of a bacterium that degrades toluene and one that degrades chlorobenzoate. In addition to developing improved strains, a great deal must be done in developing inoculation systems that assure that the desired strain(s) compete effectively and establish residence long enough to achieve the remedial objective.

Analytical Methods

Field analytical techniques for monitoring for the presence of specific degrader populations or levels of contaminants that are as easy to use as home pregnancy tests would revolutionize the environmental industry. Such methods as nucleic acid probes and monoclonal antibody tests have been developed but are not widely used because of their relatively high cost and low reliability. Are these deficiencies inherent in the technology or is further development required?

It would seem that monitoring methods that could provide direct evidence of the performance of in situ bioremediation processes would go a long way toward validating treatment effects early in the remediation process and even provide the mechanism for stimulus-response control of the process. Methods for on-line analysis of general metabolic end results, such as carbon dioxide production and oxygen consumption, are used fairly routinely. However, as the Stanford field pilot program demonstrated, additional benefit can be gained by tracking the levels of specific transient metabolic products of the biological process. These observations beg the question of whether our fundamental knowledge of biodegradative pathways can be used to suggest and/or develop similar methods for classes of contaminants in addition to petroleum hydrocarbons.

Process Innovations

A number of new technologies, biological and chemical, could be used to enhance bioremediation. With increasing knowledge of anaerobic biodegradation, it should not be long before we witness the use of this microbial process to encourage in situ biorestoration of sites contaminated with chlorinated solvents, PCBs, chlorinated pesticides, or other halogenated organics that otherwise resist microbial degradation. On purely thermodynamic grounds, it is not unreasonable to suppose that a treatment-train approach using both anaerobic and aerobic biodegradation would be the most efficient way to handle such compounds as PCE and PCBs.

A second possibility involves in situ soil flushing, a technology derived from tertiary recovery of petroleum from oil fields. Surfactant/polymer floods are used to essentially wash product or pollutants from the subsurface for above-ground recovery. Typically, this process will leave behind residual contaminants and polymer/surfactant. The potential of using in situ bioremediation to treat these residuals (biopolishing) has received minimal investigation.

CONCLUSION

Bioremediation technology has evolved over 20 years of commercial life. It started as one of the first primary treatment processes, able to address both soil and ground water contamination. It has since become an incremental technology, directed at accelerating the remediation of sites contaminated by petroleum hydrocarbons and other degradable substrates.

The evolution of bioremediation has resulted primarily from engineering work. Most advances in commercial application have been tied to improving oxygen availability. The technology has evolved from simple in-well aeration to chemical carriers such as hydrogen peroxide or nitrate and, finally, to aeration technology—soil vapor extraction and air sparging. In the course of this evolution the importance of the biological pathway has declined as physical removal processes have evolved.

The future of bioremediation lies in addressing those contaminants that are not easily extracted physically, such as PAHs, PCBs, and pesticides. This, approach, however, requires advances in the fundamental knowledge of microbial ecology and biodegradation pathways. Application of new microbial processes requires better monitoring and mathematical modeling as well as improved subsurface engineering. Such advances will lead to better understanding and use of natural or enhanced in situ bioremediation.

REFERENCES

Alvarez, P. J., and T. M. Vogel. 1991. Substrate interactions of benzene, toluene, and para-xylene during microbial degradation by pure cultures and mixed culture aquifer slurries. Applied and Environmental Microbiology 57(10):2981-2985.

American Petroleum Institute (API). 1987. Field Study of Enhanced Subsurface Biodegradation of Hydrocarbons Using Hydrogen Peroxide as an Oxygen Source. API Pub. 4448. Washington, D.C.: API.

Arvin, E., B. K. Jensen, and A. T. Gunderson. 1989. Substrate interactions during aerobic biodegradation on benzene. Applied and Environmental Microbiology 55(12):3221-3225.

Barker, J. F., G. C. Patrick, and D. Major. 1987. Natural attenuation of aromatic hydrocarbons in a shallow sand aquifer. Ground Water Monitoring Review 7(1): 64-71.

Bedard, D. L., M. L. Haberl, R. J. May, and M. J. Brennan. 1987. Evidence for novel mechanisms of polychlorinated biphenyl metabolism in Alcaligenes eutrophus H850. Applied and Environmental Microbiology 53(5):1103-1112.

Bouwer, E. J., B. E. Rittmann, and P. L. McCarty. 1981. Anaerobic degradation of halogenated 1 carbon and 2 carbon organic compounds. Environmental Science and Technology 15(5):596-599.

Bouwer, E. J., and D. G. Capone. 1988. Effects of co-occurring aromatic hydrocarbons on degradation of individual polycyclic aromatic hydrocarbons in marine sediment slurries. Applied and Environmental Microbiology 54(7):1649-1655.

Britton, L. N. 1985. Feasibility Studies on the Use of Hydrogen Peroxide to Enhance Microbial Degradation of Gasoline. API Pub. 4389. Washington, D.C.: API.

Brown, R. A., R. D. Norris, and R. L. Raymond. 1984. Oxygen transport in contaminated aquifers. Proceedings of Petroleum Hydrocarbons and Organic Chemicals in Groundwater: Prevention, Detection, and Restoration, November 5-7, Houston. Worthington, Ohio: National Well Water Association.

Brown, R. A., R. D. Norris, and R. L. Raymond. 1986. U.S. Patent 4,588,506, Stimulation of Bio-oxidation Process in Subterranean Formations. May 13.

Brown, R. A., and J. Crosbie. 1989. Oxygen Sources for In Situ Bioremediation. Greenbelt, Md.: Hazardous Materials Control Research Institute.

Brown, R. A., and R. T. Cartwright. 1990. Biotreat sludges and soils. Hydrocarbon Processing (Oct.):93-96.

Brown, R. A., and F. Jasiulewicz. 1992. Air sparging: a new model for remediation. Pollution Engineering (July):52-57.

Fathepure, B. Z., J. P. Nengu, and S. A. Boyd. 1987. Anaerobic bacteria that dechlorinate perchloroethene. Applied and Environmental Microbiology 53(11):2671-2674.

Floodgate, G. D. 1973. Theory concerning the biodegradation of oil in natural waters. In Microbial Degradation of Oil Pollutants, D. G. Ahearn and S. P. Meyers, eds., Publ. No. LSU-SG-73-01. Baton Rouge: Louisiana State University, Center for Wetland Resources.

Frankenberger, W. T., Jr., K. D. Emerson, and D. W. Turner. 1989. In situ bioremediation of an underground diesel fuel spill: a case history. Environmental Management 13(3):325-332.

Jamison, V. M., R. L. Raymond, and J. O. Hudson, Jr. 1975. Biodegradation of high-octane gasoline in groundwater. Developments in Industrial Microbiology 16:305-312.

Leadbetter, E. R., and J. W. Foster. 1959. Oxidation products formed from gaseous alkanes by the bacterium *Pseudomonas methanica*. Archives of Biochemistry and Biophysics 82:491-492.

Miller, R. N., R. E. Hinchee, C. M. Vogel, R. R. DuPont, and D. C. Downey. 1990. A field scale investigation of enhanced petroleum hydrocarbon biodegradation in the vadose zone at Tyndall AFB, Florida. In Proceedings of Petroleum Hydrocarbons and Organic Chemicals in Ground Water: Prevention, Detection, and Restoration, Oct. 31-Nov. 2, Houston. Worthington, Ohio: National Water Well Association.

Nelson, M. J. K., S. O. Montgomery, W. R. Mahaffey, and P. H. Pritchard. 1987. Biodegradation of trichloroethylene and involvement of an aromatic biodegradative pathway. Applied and Environmental Microbiology 53(5):949-954.

Nelson, M. J., J. V. Kinsella, and T. Montoya. 1990. In situ biodegradation of TCE contaminated groundwater. Environmental Progress 9(3):190-196.

Norris, R. D., and K. Dowd. 1993. Successful in situ bioremediation in a low permeability aquifer. In Bioremediation: Field Experience, P. E. Flathman, J. Exner, and D. Jerger, eds. Chelsea, Mich.: Lewis Publishers.

Quensen, J. F., J. M. Tiedje, and S. A. Boyd. 1988. Reductive dechlorination of polychlorinated biphenyls by anaerobic microorganisms from sediments. Science 242:752-754.

Raymond, R. L., V. W. Jamison, and J. O. Hudson. 1977. Beneficial stimulation of bacterial activity in groundwater containing petroleum hydrocarbons. American Institute of Chemical Engineers Symposium Series 73(166):390-404.

Reineke, W., and H. J. Knackmuss. 1990. Hybrid pathway for chlorobenzoate metabolism in pseudomonas sp. B13 derivatives. Journal of Bacteriology 142(2):467-473.

Senez, J. C., and M. Konovaltschikoff-Mazoyer. 1956. Formation d'acides gras dans les cultures de *Pseudomonas ruginosa* sur n-heptane. Comptes Rendus Hebdomadaires des Seances de l'Academie des Sciences 242:2873-2875.

Sisler, F. D., and C. E. Zobell. 1947. Microbial utilization of carcinogenic hydrocarbons. Science 106:521-522.

Tausson, W. C. 1927. Naphthalin als Kohlenstoffquelle für Bakterien. Planta 4:214-256.

Texas Research Institute, Inc. 1982. Enhancing the Microbial Degradation of Underground Gasoline by Increasing Available Oxygen, Final Report. Washington, D.C.: American Petroleum Institute.

Thornton, J. C., and W. L. Wooten. 1982. Venting for the removal of hydrocarbon vapors from gasoline contaminated soil. Journal of Environmental Science and Health A17(1):31-44.

VanLocke, R., A. M. Verlinde, W. Verstraeta, and R. DeBorger. 1979. Microbial release of oil from soil columns. Environmental Science and Technology 13(March):346-348.

Vogel, T. M., and P. L. McCarty. 1985. Biotransformation of tetrachloroethylene to trichloroethylene, dichloroethylene, vinyl chloride, and carbon dioxide under methanogenic conditions. Applied and Environmental Microbiology 49(5):1080-1083.

Wilson, J. T., and B. H. Wilson. 1985. Biotransformation of trichloroethylene in soil. Applied and Environmental Microbiology 49(1):242-243.

Wilson, J. T., and C. H. Ward. 1986. Opportunities for bioremediation of aquifers contaminated with petroleum hydrocarbons. Journal of Industrial Microbiology 27:109-116.

Wilson, J., L. Leach, J. Michalowski, S. Vandergrift, and R. Calloway. 1990. In situ reclamation of spills from underground storage tanks: new approaches for site characterization, project design, and evaluation of performance. In Proceedings: Environmental Research Conference on Groundwater Quality and Waste Disposal, May 2-4, 1989, Washington, D.C., I. P. Muraka and S. Cordle, eds. Palo Alto, Calif.: Electric Power Research Institute.

Zobell, C. E. 1973. Microbial degradation of oil: present status, problems, and perspectives, P. 15 in The Microbial Degradation of Oil Pollutants, D. G. Ahearn and S. P. Meyers, eds. Publ. No. LSU-SG-73-01. Baton Rouge: Louisiana State University, Center for Wetland Resources.

Engineering Challenges of Implementing In Situ Bioremediation

Lisa Alvarez-Cohen
University of California
Berkeley, California

SUMMARY

The use of in situ bioremediation to destroy ground water contaminants essentially requires the creation and management of a subsurface bioreactor. Physical and chemical conditions within the subsurface environment can be manipulated to optimize microbial growth by using hydrodynamic or gas-phase controls. Requisite factors for successful application of in situ bioremediation include adequate aquifer permeability; a suitable microbial population; sufficient hydrodynamic control for plume containment and delivery of required electron donors, electron acceptors, and/or nutrients; and a complete monitoring system.

Evaluating the progress of in situ bioremediation and proving that the microbes are responsible for contaminant degradation can be challenging because of the inaccessibility of the subsurface bioreactor, aquifer heterogeneities, and the wide range of potential contaminant fates. However, overlapping lines of evidence from a range of field-monitoring techniques may provide suitable indication of successful in situ bioremediation.

INTRODUCTION

In situ bioremediation involves the stimulation of microorganisms within a subsurface aquifer to degrade ground water contaminants—that is, management of a subsurface bioreactor to carry out specific biological degradations. Management of a subsurface bioreactor can be passive, involving only the monitoring of naturally occurring microbial degradations, or it can be active, involving engineered systems for the manipulation of physical and chemical conditions within the subsurface environment. Subsurface conditions can be effectively altered with hydrodynamic controls, using water as the delivery and transport system, or with gas-phase control within the vadose zone (unsaturated subsurface). Hydrodynamic controls involve manipulation of ground water flow and may include injection wells or infiltration galleries for the introduction of water to the subsurface along with production wells for ground water withdrawal. Gas-phase controls may take the form of vacuum extraction or venting systems, which may be accompanied by direct injection of gas to the vadose zone or sparging of gas into the ground water. The subsurface bioreactor or zone of biostimulation would then occur between the injection and withdrawal systems. However, the inherent heterogeneity and inaccessibility of the subsurface make in situ bioprocesses much more difficult to monitor and control than above-ground engineered systems. The following discussion addresses some of the unique engineering challenges associated with the use of in situ bioremediation to treat contaminated aquifers.

SUBSURFACE BIOREACTOR REQUIREMENTS

Before applying in situ bioremediation to a contaminated aquifer, it is necessary to evaluate the feasibility of engineering a subsurface bioreactor at the specific site to carry out the biological degradations of interest. Engineering feasibility depends on a number of factors; principal among them are aquifer permeability, heterogeneity, and geochemical characteristics, as well as the nature and distribution of the contaminants. As with above-ground bioreactors, providing appropriate environmental conditions, residence times, and substrate availability are fundamental requirements for promotion of efficient biodegradation reactions.

Subsurface Investigation

The practicality of a subsurface bioreactor depends on aquifer characteristics that can best be evaluated by a thorough site investigation, including tracer studies. A combination of existing site data and both direct and indirect measurements can be used to evaluate these characteristics and define the nature and extent of subsurface contamination. Boreholes and monitoring wells can permit direct sampling of aquifer material and ground water, which is useful for aquifer characterization and development of a three-dimensional assessment of contaminant distribution.

In determining well placement for site assessment, the cost associated with each boring must be weighed against the information lost in the spaces between borings, a balance that is highly dependent on the heterogeneity of the aquifer. It is also necessary to place wells upgradient of the contamination source to provide information on background water quality as well as downgradient for information on the location and size of the pollutant plume. Preliminary information on aquifer composition and plume location derived from remote sensing by geophysical techniques can be useful for optimizing well placement and, consequently, for decreasing the number of wells necessary for adequate monitoring (Benson et al., 1988). Additionally, gas surveys (analyses of gas samples from within the unsaturated zone) are useful for detecting volatiles that diffuse up from the water table. Such surveys may be advantageous for decreasing the number of monitoring wells since they offer additional information on location and migration of volatile plumes (LaGrega et al., 1992).

Tracer tests are additional direct measurement tools that are useful for estimating the direction and velocity of ground water flow and the hydraulic conductivity, porosity, and dispersivity of the aquifer. Tracers also facilitate estimation of contaminant residence times within the biostimulation zone, which is useful for predicting degradation efficiency.

Permeability

Adequate permeability for the transport of solutions delivering nutrients or other compounds required for stimulation of the desired microbial population within the subsurface bioreactor is essential to in situ bioremediation. Additionally, the aquifer must be sufficiently permeable that the increased microbial mass and volume will not cause extensive plugging of the aquifer pores, thus restricting further ground water movement. The proposed rule of thumb (Thomas and Ward, 1989) is that aquifers with overall hydraulic conductivities of 10^{-4} cm/s or greater would be most amenable to in situ bioremediation (10^{-4} cm/s hydraulic conductivity corresponds roughly to an intrinsic permeability of 10^{-9} cm^2 for clean water at typical subsurface temperatures). However, it has been shown that microbial growth in aquifer material can cause permeabilities to decrease by a factor of 1000 (Taylor et al., 1990). Additionally, modeling that considered microbial growth, transport, and biofilm shearing has shown that high-porosity media with widely distributed pore sizes in the small-diameter range are much more susceptible to biofouling than high-porosity media with a narrow pore size range. These results suggest that both permeability and pore size distribution must be considered in determining the feasibility of in situ bioremediation (Taylor and Jaffe, 1991).

Environmental Conditions

It is important to analyze the environmental parameters inside the intended zone of biostimulation that could exert significant impact on microbial growth and degradation potential. Microbial metabolism is substantially affected by temperature: the metabolism of subsurface populations tends to accelerate with increased subsurface temperatures within typical (nongeothermal) ranges. Although temperatures within the top 10 m of the subsurface may fluctuate seasonally, subsurface temperatures down to 100 m typically remain within 1° to 2°C of the mean annual surface temperature (Freeze and Cherry, 1979), suggesting that bioremediation within the subsurface would occur more quickly in temperate climates (Lee et al., 1988).

Additional factors that may limit microbial activity have been summarized elsewhere (Ghiorse and Balkwill, 1985; Ghiorse and Wilson, 1988). They include pH values outside the range of neutral ($pH<6$, $pH>8$), desiccating moisture conditions, and extreme redox (reduction-oxidation) potentials. Each of these factors may be mitigated or controlled within a desired range with varying levels of success using hydrodynamic controls.

Monitoring

Monitoring of ground water and aquifer conditions over time is necessary for assessing activity within the subsurface bioreactor and evaluating the progress of the bioremediation. Monitoring wells typically are installed between the injection and production wells so as to detect microbial growth and contaminant degradation within the biostimulation zone. Additional wells installed upgradient from the contamination provide background characterization data, while wells installed beyond the downgradient contaminant boundaries are useful for detection of plume expansion or migration. Factors that should be analyzed in the monitoring samples include contaminant concentration, microbial numbers, electron donor and acceptor concentrations, oxygen demand, degradation products, pH, and major ion concentrations. Sample collection and handling procedures have been summarized elsewhere (Barcelona et al., 1990).

AQUIFER PREPARATION

Before applying in situ bioremediation, the source of contamination must be detected and mitigated, major accumulations of free product must be removed, and mechanisms for plume containment must be installed. Contaminants entering the subsurface partition into different phases due to sorption, volatilization, and dissolution processes. Contaminant partitioning impedes pump-and-treat removal methods and may decrease the contaminant's availability to microbial degradation.

Source and Free Product Removal

The first step in most aquifer remediation efforts is removal or mitigation of the contaminant source: excavation of leaking underground storage tanks, plugging or repairing leaking surface impoundments or landfill liners, restricting intentional or unintentional land application, and similar measures. Removal typically is achieved by excavating the most contaminated surface soils and drilling wells to pump out the most concentrated source material. Liquid contaminants, which often exist as nonaqueous-phase liquids, or free product, are drawn into the subsurface by gravity and capillary action within the porous media. Free product that is lighter than water, such as petroleum hydrocarbons, tends to migrate downward through the unsaturated zone until hitting an impermeable layer or the water table, where it spreads laterally. Free product that is denser than water, such as many of the chlorinated solvents, would continue to migrate downward through the water table and saturated zone until reaching an impermeable barrier. Free product can be removed from an aquifer by direct pumping using production wells alone or combinations of injection and production wells to cause directional migration of the floating or sinking product. However, most pumping strategies will be capable of removing only part of the free product from the subsurface, leaving the remainder of the organic material trapped in pores as residual free product, dissolved in the surrounding ground water as a contaminant plume, sorbed onto the solid subsurface material, or volatilized into the gas-filled pores of the unsaturated zone. Additionally, the location and removal of sinking free product typically represent much more of an engineering challenge than that of floating product, often resulting in low recovery efficiency.

Plume Containment

In situ bioremediation typically requires times on the order of months to years to reduce contaminants to acceptable levels. During that time the contaminants must not be allowed to spread outside the bioremediation zone and thereby escape treatment.

A contaminant plume can be contained by physical or hydrodynamic controls or a combination of both. Physical controls include low-permeability vertical walls installed to physically block the transport of the plume and/or to inhibit the flow of clean ground water into the contaminated zone. The most commonly used physical containment barrier is the slurry trench wall, which typically is composed of a mixture of bentonite and soil or bentonite and cement. A slurry wall keyed into a confining impermeable layer can significantly decrease localized ground water flow and lengthen the ground water flow path. Grout curtains, vibrating beam walls, and synthetic sheet curtains are also used on a limited basis for physical containment. Physical barriers are most effective with shallow aquifers underlayed by a solid confining layer of bedrock or clay (LaGrega et al., 1992).

Hydrodynamic controls are used alone or in conjunction with physical controls. They are especially suited for use with in situ bioremediation since biostimulation amendments could be added with the control water. Hydrodynamic controls typically consist of combinations of injection and extraction wells and/or infiltration galleries that manipulate ground water flow in order to prevent undesirable plume movement. Wells are situated so that their radii of influence (area of water drawdown or mounding) overlap, allowing control of water within the entire treatment zone as well as effective manipulation of the level of the water table. Radii of influence are computed by iterative application of steady pumping rates with drawdown equations appropriate to the specific aquifer conditions. Plume direction, shape, and migration speed can each be effectively manipulated by hydrodynamic controls, which regulate the detention time and amendment delivery within the biostimulation zone (Barcelona et al., 1990; Knox et al., 1986).

IN SITU BIOSTIMULATION

Stimulation of microbial populations within a subsurface bioreactor requires an appropriate carbon source; electron donors/acceptors for energy production; and inorganic nutrients such as nitrogen, phosphorus, and some trace metals. Also required are proper conditions within the aquifer, such as appropriate pH, temperature, moisture content, and redox potential.

Indigenous microbial populations from many aquifers have been shown to be capable of degrading a wide range of organic contaminants, obviating the need for introduction of exogenous cultures in most bioremediation applications (Lee et al., 1988). However, in the absence of appropriate indigenous strains, introduction of laboratory-enriched populations or even genetically engineered microorganisms may be possible.

To detect the presence of microbial populations capable of degrading the contaminant of interest, laboratory feasibility studies should be conducted. In these studies an aseptically collected sample of subsurface material is exposed to the contaminant of interest under simulated aquifer conditions and is analyzed for contaminant degradation and the concurrent appearance of degradation products. Aseptic aquifer samples are collected by withdrawing an uncontaminated section of a drilling core using a sterile paring device (Lee et al., 1988). Radiotracers may also be used as an analytical tool to confirm contaminant degradation and to trace the degradation sequence. Additional information that may be collected from feasibility studies includes the range of nontoxic contaminant concentrations, nutrient and electron donor/acceptor requirements for optimizing cellular growth and contaminant degradation, and estimates of microbial acclimation periods and growth rates. Adequate detention time for the contaminants within the biostimulation zone must be maintained to allow the degradation reaction to reduce the contaminant concentrations to the desired levels, and acclimation periods may be necessary before the indigenous population becomes capable of carrying out the degradation reaction (Wilson et al., 1985).

Role of Nutrients, Electron Donors, Acceptors

Microorganisms produce energy by moving electrons between an electron donor and an electron acceptor. These reactions can be carried out aerobically, using oxygen as the electron acceptor, or anaerobically, using nitrate, sulfate, carbon dioxide, or other oxidized species as the electron acceptor. Many organic contaminants can be used as a primary substrate for microbial metabolism, in which case the contaminant serves as an electron donor and sometimes also as the major carbon source for the microbial cells. Therefore, some degradation reactions produce energy and usable carbon, resulting in microbial growth. Hence, bioremediation of a primary substrate has a built-in termination mechanism: as the contaminant/substrate is consumed, resulting in depleted concentrations, microbial growth slows and ceases. The hydrocarbons in gasoline and other petroleum derivatives can be aerobically degraded and used as a primary substrate for growth by a wide range of naturally occurring microorganisms (Armstrong et al., 1991; Ridgway et al., 1990). Bioremediation of gasoline-contaminated aquifers by indigenous microflora has been successfully implemented a number of times (Jamison et al., 1976; Lee et al., 1988; Thomas and Ward, 1989), although researchers have sometimes reported that the indigenous microbial population required a period of adaptation before degradation commenced (Armstrong et al., 1991).

Alternate Substrates

Because of the nature of contaminant partitioning within the subsurface, contaminants may be transported through the aquifer in dilute ground water plumes at concentrations that provide insufficient energy and/or carbon to support microbial growth. Additionally, since aquifers may be used as drinking water sources, the allowable levels of many ground water contaminants are set in the range of micrograms per liter, requiring reduction of contaminants to concentrations lower than those required for microbial reproduction. Under these circumstances, it may be necessary to supply an alternate substrate for the microorganisms in order to promote degradation by *secondary metabolism*. Similarly, contaminants that do not benefit microorganisms by providing energy or carbon can sometimes be degraded in the presence of an alternate microbial substrate by *cometabolism*. Some examples of cometabolic degradations are those catalyzed by the mono- or dioxygenase enzymes of methane-, propane-, or toluene-oxidizing bacteria, which use the contaminant as electron donor and oxygen as electron acceptor, or those carried out by anaerobic bacteria capable of using the contaminant as the electron acceptor. A wide range of chlorinated solvents, including trichloroethylene and vinyl chloride, have been shown to be cometabolically degraded by methane- and toluene-oxidizing bacteria. For the application of either secondary metabolism or cometabolism, use of an alternate substrate presents an additional engineering challenge and potential limit on the reaction rate. However, the use of an alternate substrate also enables the degradation reaction to be maintained at contaminant concentrations below those required to support microbial growth and much lower than those possible for degradations in which the contaminant is used by the microorganisms as a primary substrate. Therefore, alternate substrates can increase the potential for attaining contaminant removal to regulatory levels.

Nutrient Delivery

Nutrients typically are delivered by controlling ground water flow using injection wells or infiltration galleries coupled with downstream production wells. In the most common configuration, ground water withdrawn from production wells downgradient from the biostimulation zone is amended with the nutrients required for biostimulation, treated if necessary to remove contaminants, and reintroduced to the aquifer upgradient of the biostimulation zone using the injection wells or infiltration galleries. Water from an external source is required if the flow of withdrawn water is insufficient to control the subsurface flow or if it is infeasible to reinject the withdrawn ground water. The rate of nutrient delivery to the biostimulation zone, therefore, is often limited by the solubility of the nutrients in water and the reinjection flow rate.

Alternately, gaseous nutrients or substrates such as oxygen or methane may be delivered to the biostimulation zone by sparging, the direct injection of gas into the saturated aquifer to effect in situ dissolution of the gas into solution. However, mobilization of volatile contaminants into the gas phase may necessitate additional gas-phase controls.

When the limiting nutrients for microbial growth are added to the subsurface, excessive microbial growth may occur around the injection zone, causing significant plugging of the permeable media and limiting the reinjection flow. Innovative methods for discouraging well plugging while promoting dispersed microbial growth throughout the zone of contamination are required. One such method, which has been shown in field studies to reduce localized plugging associated with cometabolic bioremediation, is alternating pulses of electron donor and electron acceptor in the reinjection water. Since both electron donor and acceptor are required for microbial metabolism, advective and dispersive processes within the aquifer must mix the nutrients before conditions promote microbial growth, causing cells to grow dispersed throughout the aquifer and producing a large biostimulation zone (Semprini et al., 1990).

Oxygen, Air, Hydrogen Peroxide

Because of the low solubility of oxygen in water, the major kinetic limitation on aerobic bioremediation reactions is often the availability of oxygen (Lee et al., 1988; Thomas and Ward, 1989). This is especially the case with high BOD (biological oxygen demand) compounds such as petroleum hydrocarbons. Air sparging of water can supply 8 mg/l dissolved oxygen, while sparging with pure oxygen can deliver 40 mg/l and with hydrogen peroxide more than 100 mg/l oxygen. Therefore, while air sparging is the simplest and most common oxygen delivery technique, the use of oxygen or hydrogen peroxide may speed the bioremediation process and decrease the pumping required. However, in some cases the increased cost and potential explosion hazard associated with pure oxygen may more than offset its increased delivery efficiency.

Application of hydrogen peroxide to in situ bioremediation is limited by its toxicity to microbes and its potential for causing aquifer plugging. Two molecules of peroxide are required to produce one molecule of oxygen:

$$2H_2O_2 = O_2 + 2H_2O$$

Although this reaction can be catalyzed by microorganisms, it is also catalyzed by naturally occurring compounds in aquifer material. The highly reactive nature of hydrogen peroxide results in chemical oxidations of organic and inorganic compounds, producing precipitants that may contribute to aquifer plugging and may decrease the oxygen-carrying capacity of the water (Spain et al., 1989). Additionally, both metal-catalyzed and microbially induced decomposition of hydrogen peroxide may produce oxygen at concentrations above water saturation, causing bubbles to form and further decreasing aquifer permeability (Morgan and Watkinson, 1992; Pardieck et al., 1992). To mitigate undesirable peroxide reactions, phosphate is sometimes added before the peroxide to precipitate iron and thereby diminish the metal-catalyzed decomposition. Additional chelating agents have been shown to decrease metal-catalyzed decomposition; however, in a biologically active zone the majority of peroxide decomposition would be expected to be biologically induced (Morgan and Watkinson, 1992). Therefore, the dual actions of precipitation of oxidation products and bubble formation typically limit the practical concentration for addition of hydrogen peroxide in ground waters to 100 mg/l or less (corresponding to 47 mg/l oxygen or less).

The reactivity of hydrogen peroxide in aquifers can be expected to vary considerably from site to site and may not result in significant plugging problems in very highly permeable soils and gravels. However, peroxide is also capable of causing mobilization of undesirable metals such as lead and antimony, producing additional ground water contamination. Therefore, it is important to do laboratory feasibility studies before using hydrogen peroxide in an aquifer since the range of potential adverse reactions is so great.

Finally, hydrogen peroxide can be significantly toxic to microbial populations at relatively low concentrations. The toxicity of the compound reportedly is species specific and depends on cell density (Lee et al., 1988; Pardieck et al., 1992). Significant toxicity has been reported for peroxide concentrations as low as 3 mg/l, whereas other studies have shown addition of 270 mg/l to exert no adverse effects. Additionally, acclimation of microorganisms to slowly increasing concentrations of peroxide has been reported with successful additions greater than 2000 mg/l (Pardieck et al., 1992). Again, laboratory feasibility studies will be necessary to determine the tolerance range of the indigenous microbial population.

The toxic effects of hydrogen peroxide on microbes have a side benefit that can be exploited through careful control of the injection stream. A relatively high concentration of hydrogen peroxide in the injection water may be useful for controlling the growth of biofilms within the immediate vicinity of the injection wells or infiltration galleries, providing the peroxide concentration decreases sufficiently with migration through the aquifer to preclude toxic microbial effects within the biostimulation zone, bubble formation, and precipitation of oxides.

Inorganic Nutrients

Studies of the effects of inorganic nutrient addition on bioremediation rates have yielded varying results. Experiments using nitrogen and phosphorous amendments have shown that they enhanced metabolic activities in some aquifer samples while having no significant effect in other samples from the same aquifer (Swindoll et al., 1988). Others reported that addition of nitrogen and phosphorus enhanced in situ gasoline degradation (Jamison et al., 1976). A series of microcosm and field studies suggest that enhancement of biodegradation by addition of inorganic nutrients is extremely case specific (Baker and Herson, 1990). It has been further suggested that not only are inorganic nutrients not always effective but in some cases they inhibit microbial degradation (Morgan and Watkinson, 1992). Morgan and Watkinson have also shown that phosphate addition in combination with hydrogen peroxide may cause precipitation of insoluble salts during migration through the aquifer, decreasing the permeability within the biostimulation zone.

Hence, the evidence indicates that chemical analysis is not adequate for predicting necessary nutrient amendments. Laboratory or field studies with aquifer material are needed to determine nutrient amendments required to promote maximum cell growth and contaminant degradation and to predict potential reactivity between the aquifer material and the amendments.

Microbial Introduction

Although in situ bioremediation is a developing technology and is currently the subject of many research studies, little is known about the movement of microbial cells through the subsurface matrix or the feasibility of introducing a stable mixed population of organisms into a contaminated site for the purpose of remediation (Thomas and Ward, 1989). Microbial populations suitable for introduction to a contaminated aquifer may have been selectively enriched in the laboratory or genetically engineered to carry out specific degradation reactions, to resist certain toxic effects, or to grow preferentially under specific environmental conditions. However, while laboratory-enriched populations may be added to the subsurface with little regulatory concern, the introduction of genetically engineered cultures currently is not allowed in the United States.

Successful microbial introduction requires a range of factors: (1) the population must be capable of surviving and growing in the new environment; (2) the microorganisms must retain their degradative abilities under the new conditions; (3) the organisms must come in contact with the contaminants; and (4) the electron donors/acceptors and nutrients necessary for microbial growth and contaminant degradation must be made available to the population (Thomas and Ward, 1989). Once the microorganisms are injected into the aquifer, there must be some mechanism for dispersing them throughout the biostimulation zone before they attach to the solid matrix and carry out the degradation reaction of interest. Cell transport within porous media is highly dependent on the characteristics of both the solid media and the microbial cells. Experiments have shown that the conditions that best promote microbial transport in porous media include (in order of their importance) highly permeable media, ground water of low ionic strength, and small-diameter cells (Fontes et al., 1991). To date, there has been little convincing evidence for successful in situ remediation of aquifers resulting from introduced microbial populations.

DETERMINING THE SUCCESS OF IN SITU BIOREMEDIATION

Perhaps the biggest challenges associated with managing a subsurface bioreactor for in situ bioremediation are evaluating its progress in the field and determining when sufficient contaminant destruction has occurred to warrant discontinuation of the biostimulation. For in situ bioremediation to be deemed successful, it must be shown that the mass of contaminant in the aquifer has been decreased to desired levels and that the microbial population caused the decrease. Factors such as the heterogeneous nature and inaccessibility of subsurface aquifers, together with competing concurrent processes that affect the form and location of the contaminant (such as volatilization, sorption, dissolution, migration, and dilution), all conspire to confound mass balance analyses. Therefore, specific documentation of the successful application of in situ bioremediation for the destruction of aquifer contaminants is extremely rare. It has been asserted that true proof of in situ bioremediation requires convergent lines of independent evidence of microbial degradation in the field (Madsen, 1991). These include diminished contaminant concentrations within both the horizontal and the vertical dimensions of the plume; increased microbial growth on the contaminant of interest in samples taken from the biostimulation zone; and detection of metabolic products coupled with diminished substrate concentrations (Madsen, 1991). These types of evidence may not be readily obtainable because of the complexity of the concurrent physical, chemical, and biological processes involved, aquifer heterogeneities, and site-monitoring limitations. However, a range of innovative sampling techniques may be incorporated into the field-monitoring methods in order to measure and quantify in situ bioremediation and provide a preponderance of supporting evidence.

Field-Monitoring Methods

The following is a sampling of the diverse field methods that have been applied with varying degrees of success to determine and quantify the success of in situ bioremediation:

1. Two field studies in which in situ bioremediation was based on cometabolism were performed within an initially uncontaminated confined aquifer (Semprini et al., 1990, 1991). A series of tracer studies were conducted using bromide to determine flow characteristics and capture efficiency and chlorinated organics to determine sorption and retardation factors and to evaluate the initial degradation potential within the aquifer. Afterward the aquifer was enriched for methane-oxidizing microorganisms that aerobically degrade chlorinated organics in one field study and for denitrifying organisms that anaerobically degrade chlorinated organics in the other. Compari-

sons of chlorinated organic concentrations recovered before and after biostimulation, and experiments in which the required electron donors and/or acceptors were eliminated, indicated that in situ bioremediation was successful in both field studies. The transport and degradation of the organics were quantified and the reaction kinetics were calculated by fitting models to the experimental data. The extensive instrumentation and monitoring facilities used in these studies provided much more accurate mass balances than could be expected for typical field applications.

2. Natural gradient tracer tests were used to compare methane oxidation activities in pristine and sewage-contaminated aquifers (Smith et al., 1991). Two inert ion tracers, chloride and bromide, and an inert dissolved gas tracer, hexafluoroethane, were used to determine advective and diffusive ground water characteristics. Methane breakthrough curves were measured at both sites, and methane oxidation was estimated from differences between tracer and methane recoveries. Methane oxidation was confirmed by injecting carbon-13-labeled methane and by recovering carbon-13-labeled carbon dioxide. Quantification of the methane degradation was possible, and a one-dimensional transport-and-decay model was used with the field data to determine kinetic degradation parameters.

3. Push/pull tests were used in a field study to determine whether oxygen addition was enhancing the subsurface degradation of polynuclear aromatics (Borden et al., 1989). Ground water was extracted from the aquifer, mixed with aromatics and chloride tracer, oxygenated or deoxygenated, and then rapidly reinjected into the bioactive zone. Samples from the reinjection well were analyzed periodically to determine oxygen, aromatics, chloride, and conductivity. The tracer was used to determine the recapture efficiency and to help compensate for dilution factors. While a comparison of results from the oxygenated and deoxygenated tests suggested that oxygen enhanced aromatic degradation, thus proving successful bioremediation, data from the push/pull test alone were not sufficient to quantify the degradation.

4. Indirect evidence for remediation of coal tar constituents was collected by conducting laboratory comparisons of degradation activities and microbial distributions in contaminated and pristine core samples (Madsen et al., 1991). Increased numbers of contaminant-degrading microorganisms in the contaminated cores, coupled with increased populations of protozoans within the contaminant plume, provided evidence for in situ bioremediation but did not permit quantification of the degradation.

5. In situ bioremediation of jet fuel was qualitatively demon-

strated in an actively vented region of the unsaturated zone by comparing ratios of stable isotopes (carbon-13/carbon-12) associated with carbon dioxide from atmospheric and vented gas samples (Hinchee et al., 1991). Active jet fuel degradation was confirmed by assuming that higher isotope ratios indicated atmospheric or plant respiratory origin of the carbon dioxide, while lower ratios indicated petroleum hydrocarbon degradation. However, while this approach may offer qualitative evidence for in situ bioremediation, the results are nonquantifiable.

CONCLUSIONS

In situ bioremediation is the management of a subsurface bioreactor to carry out specific biological degradations of ground water contaminants. Successful implementation should include a thorough aquifer characterization, removal of contaminant source and free product, plume containment, laboratory feasibility studies, installation and operation of biostimulation controls, and continuous monitoring.

Although proving the success of in situ bioremediation is challenging, a variety of field methods can be used to provide adequately convincing evidence of success. Quantification of in situ bioremediation, however, is much more difficult, requiring mass balances that may be achievable only under the most controlled circumstances.

ACKNOWLEDGMENTS

This work was sponsored in part by the National Institute of Environmental Health Sciences under grants P42-ES047905 and P42-ES04705 and by the U.S. Department of Energy Junior Faculty Research Award Program administered by Oak Ridge Associated Universities.

REFERENCES

Armstrong, A. Q., R. E. Hodsen, H. M. Hwang, and D. L. Lewis. 1991. Environmental factors affecting toluene degradation in ground water at a hazardous waste site. Environmental Toxicology and Chemistry 10:147-158.

Baker, K. H., and D. S. Herson. 1990. In situ bioremediation of contaminated aquifers and subsurface soils. Geomicrobiology Journal 8:133-145.

Barcelona, M., W. Wehrmann, J. F. Keely, and W. A. Pettyjohn. 1990. Contamination of Groundwater: Prevention, Assessment, Restoration. Park Ridge, N.J.: Noyes Data Corp.

Benson, R. C., M. Turner, P. Turner, and W. Vogelsong. 1988. In situ, time-series measurements for long-term groundwater monitoring. Pp. 58-72 in Ground-Water Contamination: Field Methods, A. G. Collins and A. I. Johnson, eds. Philadelphia: ASTM Publications.

Borden, R. C., M. D. Lee, J. M. Thomas, P. B. Bedient, and C. H. Ward. 1989. In situ measurement and numerical simulation of oxygen limited biotransformation. Groundwater Monitoring Review 9:83-91.

Fontes, D. E., A. L. Mills, G. M. Hornberger, and J. S. Herman. 1991. Physical and chemical factors influencing transport of microorganisms through porous media. Applied and Environmental Microbiology 57:2473-2481.

Freeze, R. A., and J. A. Cherry. 1979. Groundwater. Englewood Cliffs, N.J.: Prentice-Hall.

Ghiorse, W. C., and D. L. Balkwill. 1985. Microbial characterization of subsurface environments. In Groundwater Quality, C. H. Ward, W. Giger, and P. L. McCarty, eds. New York: John Wiley & Sons.

Ghiorse, W. C., and J. T. Wilson. 1988. Microbial ecology of terrestrial subsurface. Advances in Applied Microbiology 33:107-172.

Hinchee, R. E., D. C. Downey, R. R. Dupont, P. K. Aggarwal, and R. N. Miller. 1991. Enhancing biodegradation of petroleum hydrocarbons through soil venting. Journal of Hazardous Materials 27:315-325.

Jamison, V. W., R. L. Raymond, and J. O. Hudson. 1976. Biodegradation of high-octane gasoline in groundwater. Developments in Industrial Microbiology 16:305-312.

Knox, R. C., L. W. Canter, D. G. Kincannon, E. L. Stover, and C. H. Ward. 1986. Aquifer Restoration: State of the Art. Park Ridge, N.J.: Noyes Publications.

LaGrega, M. D., P. L. Buckingham, and J. C. Evans. 1992. The ERM Group's Hazardous Waste Management. New York: McGraw-Hill.

Lee, M. D., J. M. Thomas, R. C. Borden, P. B. Bedient, C. H. Ward, and J. T. Wilson. 1988. Biorestoration of aquifers contaminated with organic compounds. Chemical Rubber Company Critical Reviews in Environmental Control 18:29-89.

Madsen, E. L. 1991. Determining in situ biodegradation. Environmental Science and Technology 25(10):1663-1673.

Madsen, E. L., J. L. Sinclair, and W. C. Ghiorse. 1991. In situ biodegradation: microbiological patterns in a contaminated aquifer. Science 252:830-833.

Morgan, P., and R. J. Watkinson. 1992. Factors limiting the supply and efficiency of nutrient and oxygen supplements for the in situ biotreatment of contaminated soil and groundwater. Water Research 26(1):73-78.

Pardieck, D. L., E. J. Bouwer, and A. T. Stone. 1992. Hydrogen peroxide use to increase oxidant capacity for in situ bioremediation of contaminated soils and aquifers: a review. Journal of Contaminant Hydrology 9:221-242.

Ridgway, H. F., J. Safarik, D. Phipps, and D. Clark. 1990. Identification and catabolic activity of well-derived gasoline degrading bacteria from a contaminated aquifer. Applied and Environmental Microbiology 56:3565-3575.

Semprini, L., P. V. Roberts, G. D. Hopkins, and P. L. McCarty. 1990. A field evaluation of in-situ biodegradation of chlorinated ethenes. Groundwater 28:715-727.

Semprini, L., G. D. Hopkins, P. V. Roberts, and P. L. McCarty. 1991. In situ biotransformation of carbon tetrachloride, freon-113, freon-11, and 1,1,1-TCA under anoxic conditions. Pp. 41-58 in On-Site Bioreclamation, R. E. Hinchee and R. F. Olfenbuttel, eds. Boston: Butterworth.

Smith, R. L., B. L. Howles, and S. P. Garabedian. 1991. In situ measurement of methane oxidation in groundwater by using natural-gradient tracer tests. Applied and Environmental Microbiology 57(7):1997-2004.

Spain, J. C., J. D. Milligan, D. C. Downey, and J. K. Slaughter. 1989. Excessive bacterial decomposition of H_2O_2 during enhanced biodegradation. Groundwater 27:163-167.

Swindoll, C. M., C. M. Aelion, and F. K. Pfaender. 1988. Influence of inorganic and organic nutrients on aerobic biodegradation and on the adaptation response of subsurface microbial communities. Applied and Environmental Microbiology 54(1):212-217.

Taylor, S. W., and P. R. Jaffe. 1991. Enhanced in-situ biodegradation and aquifer permeability reduction. Journal of Environmental Engineering 117(1):25-46.

Taylor, S. W., P. C. D. Milly, and P. R. Jaffe. 1990. Biofilm growth and the related changes in the physical properties of a porous medium; permeability. Water Resources Research 26(9):2161-2169.

Thomas, J. M., and C. H. Ward. 1989. In situ biorestoration of organic contaminants in the subsurface. Environmental Science and Technology 23(7):760-766.

Wilson, J. T., J. F. McNabb, J. W. Cochran, T. H. Wang, M. B. Tomson, and P. B. Bedient. 1985. Influence of microbial adaptation on the fate of organic pollutants in groundwater. Environmental Toxicology and Chemistry 4:721.

Modeling In Situ Bioremediation

Philip B. Bedient and Hanadi S. Rifai
Rice University
Houston, Texas

SUMMARY

The problem of quantifying biodegradation of subsurface pollutants can be addressed by using models that combine physical, chemical, and biological processes. Developing such models is difficult, however, for reasons that include the lack of field data on biodegradation and the lack of numerical schemes that accurately simulate the relevant processes. This paper reviews modeling efforts, including BIOPLUME II.

INTRODUCTION

One of the aquifer remediation methods that has been gaining more widespread attention recently is bioremediation, the treatment of subsurface pollutants by stimulating the growth of native microbial populations. The purpose is to biodegrade complex hydrocarbon pollutants into simple carbon dioxide and water. The technology is not novel; biodegradation of organic contaminants has been recognized and utilized in the wastewater treatment process for years.

Bioremediation is not without its problems, however. The most important are the lack of well-documented field demonstrations, preferably quantitative, of the effectiveness of the technology and its long-term effects, if any, on ground water systems. Other problems include the possibility that the biodegradation process will generate undesirable intermediate compounds that are more persistent in the environment than the parent compounds.

MODELING BIODEGRADATION PROCESSES

The problem of quantifying biodegradation in the subsurface can be addressed by using models that combine physical, chemical, and biological processes. Developing such models is not simple, however, because of the complex nature of microbial kinetics, the limitations of computer resources, the lack of field data on biodegradation, and the lack of robust numerical schemes that can simulate the physical, chemical, and biological processes accurately. Several researchers have developed ground water biodegradation models. The main approaches used for modeling biodegradation kinetics are:

- first-order degradation models,
- biofilm models (including kinetic expressions),
- instantaneous reaction models, and
- dual-substrate Monod models.

These are described in more detail in the next section. A more thorough discussion of models can be found in an earlier National Research Council report (National Research Council, 1990).

Previous Modeling Efforts

McCarty et al. (1981) modeled the biodegradation process using biofilm kinetics. They assumed that substrate concentration within the biofilm changes only in the direction normal to the surface of the biofilm and that all the required nutrients except the rate-limiting substrate are in excess. The model employs three basic processes: mass transport from the bulk liquid, biodecomposition within the biofilm, and biofilm growth and decay. The authors evaluated the applicability of the biofilm model to aerobic subsurface biodegradation using a laboratory column filled with glass beads. The experimental data and the model predictions were relatively consistent.

Kissel et al. (1984) developed differential equations describing mass balances on solutes and mass fractions in a mixed-culture biological film within a completely mixed reactor. The model incorporates external mass transport effects, Monod kinetics with internal determination of limiting electron donor or acceptor, competitive and sequential reactions, and multiple active and inert biological fractions that vary spatially. Results of hypothetical simulations involving competition between heterotrophs that derive energy from an organic solute and autotrophs that derive energy from ammonia and nitrite were presented.

Molz et al. (1986) and Widdowson et al. (1987) presented one- and two-dimensional models for aerobic biodegradation of organic contaminants in ground water coupled with advective and dispersive transport. A microcolony approach was used in the modeling effort: microcolonies of bacteria are represented as disks of uniform radius and thickness attached to aquifer sediments. Associated with each colony was a boundary layer of a given thickness across which substrate and oxygen are transported by diffusion to the colonies. The authors' results indicated that biodegradation would be expected to have a major effect on contaminant transport when proper conditions for growth exist. Simulations of two-dimensional transport suggested that under aerobic conditions microbial degradation reduces the substrate concentration profile along longitudinal sections of the plume and retards the lateral spread of the plume. Anaerobic conditions developed in the center of the plume because of microbial consumption and limited oxygen diffusion into the plume's interior.

Widdowson et al. (1988) extended their previous work to simulate oxygen- and/or nitrate-based respiration. Basic assumptions incorporated into the model include a simulated particle-bound microbial population comprised of heterotrophic facultative bacteria in which metabolism is controlled by lack of an organic carbon electron donor source (substrate), an electron acceptor (oxygen and/or nitrate), a mineral nutrient (ammonium), or all three simultaneously.

Srinivasan and Mercer (1988) presented a one-dimensional, finite difference model for simulating biodegradation and sorption processes in saturated porous media. The model is formulated to accommodate a variety of boundary conditions and process theories. Aerobic biodegradation was modeled using a modified Monod function; anaerobic biodegradation was modeled using Michaelis-Menten kinetics. In addition, first-order degradation was allowed for both substances. Sorption was incorporated using linear, Freundlich, or Langmuir equilibrium isotherms for either substance.

MacQuarrie and Sudicky (1990) used the model developed by MacQuarrie et al. (1990) to examine plume behavior in uniform and random flow fields. In uniform ground water flow, a plume originating from a high-concentration source will experience more spreading and slower normalized mass loss than a plume from a source of lower initial concentration because dissolved oxygen is more quickly depleted. Large ground water velocities produced increases in the rate of organic solute mass loss because of increased mechanical mixing of the organic plume with oxygenated ground water.

Development and Application of BIOPLUME

Borden and Bedient (1986) developed the first version of the BIOPLUME model. They developed a system of equations to simulate the simultaneous growth, decay, and transport of microorganisms combined with the transport and removal of hydrocarbons and oxygen. Simulation results indicated that any available oxygen in the region near the hydrocarbon source will be rapidly consumed. In the body of the hydrocarbon plume, oxygen transport will be rate limiting and the consumption of oxygen and hydrocarbon can be approximated as an instantaneous reaction. The major sources of oxygen, this research concluded, are transverse mixing, advective fluxes, and vertical exchange with the unsaturated zone.

Rifai et al. (1987, 1988) expanded and extended the original BIOPLUME and developed a numerical version of the biodegradation model (BIOPLUME II) by modifying the U.S. Geological Survey (USGS) two-dimensional method of characteristics model (Konikow and Bredehoeft, 1978). The basic concept used in developing BIOPLUME II includes the use of a dual-particle mover procedure to simulate the transport of oxygen and contaminants in the subsurface.

Biodegradation of the contaminants is approximated by the instantaneous reaction model. The ratio of oxygen to dissolved contaminants consumed by the reaction is determined from an appropriate stoichiometric model (assuming complete mineralization). In general, the transport equation is solved twice at every time step to calculate the oxygen and contaminant distribution:

$$\frac{\partial (Cb)}{\partial t} = \frac{1}{R_c}\left[\frac{\partial}{\partial x_i}\left(bD_{ij}\frac{\partial C}{\partial x_j}\right) - \frac{\partial}{\partial x_i}(bCV_i)\right] - \frac{C'W}{n}$$

$$\frac{\partial (Ob)}{\partial t} = \left[\frac{\partial}{\partial x_i}\left(bD_{ij}\frac{\partial O}{\partial x_j}\right) - \frac{\partial}{\partial x_i}(bOV_i)\right] - \frac{O'W}{n}$$

where C and O are the concentration of contaminant and oxygen, respectively; C' and O' are the concentration of contaminant and oxygen in a source or sink fluid; n is the effective porosity; b is the saturated thickness; t is time; x_i and x_j are Cartesian coordinates; W is the volume flux per unit area; V_i is the seepage velocity in the direction of x_i; R_c is the retardation factor for the contaminant; and D_{ij} is the coefficient of hydrodynamic dispersion.

The BIOPLUME II model simulates dissolved contaminant concentrations vertically averaged over the thickness of the aquifer. The two plumes are combined using the principle of superposition to simulate the instantaneous reaction between oxygen and the contaminants, and the decrease in contaminant and oxygen concentrations is calculated from:

$$DC_{RC} = O/F;\ O = 0 \quad \text{where } C > O/F$$

$$DC_{RO} = C \bullet F;\ C = 0 \quad \text{where } O > C \bullet F$$

where DC_{RC} and DC_{RO} are the calculated changes in the concentrations of contaminant and oxygen, respectively, caused by biodegradation, and F is the ratio of oxygen to contaminant consumed.

The only input parameters to BIOPLUME II that are required to simulate biodegradation are the amount of dissolved oxygen in the aquifer prior to contamination and the oxygen demand of the contaminant determined from a stoichiometric relationship. Other parameters are the same as would be required to run the standard USGS model in two dimensions (Konikow and Bredehoeft, 1978).

Borden et al. (1986) used the first version of the BIOPLUME model to simulate biodegradation of polycyclic aromatic hydrocarbons at the Conroe Superfund site in Texas. Oxygen exchange with the unsaturated zone was simulated as a first-order decay in hydrocarbon concentration. The loss of hydrocarbon because of horizontal mixing with oxygenated ground water and resulting biodegradation was simulated by generating oxygen and hydrocarbon distributions independently and then combining them by superposition. Simulated oxygen and hydrocarbon concentrations closely matched the observed values at the Conroe site.

Rifai et al. (1988) used BIOPLUME II to model biodegradation of aviation fuel at the U.S. Coast Guard Station, Traverse City, Michigan (Figure 1). Vertically averaged plume data came from 25 wells. The modeling results along the centerline of the contaminant plume were good, and the BIOPLUME II results matched the field observations except in an area between monitoring well M30 and the pumping wells.

Chiang et al. (1989) used BIOPLUME II to characterize hydrocarbon biodegradation in a shallow aquifer. They measured the soluble hydrocarbon concentrations and dissolved oxygen levels in monitoring wells. Results from 10 sampling periods over 3 years showed a significant reduction in total benzene mass with time. The natural attenuation rate was calculated to be 0.95 percent per day. Spatial relationships between dissolved oxygen and total benzene, toluene,

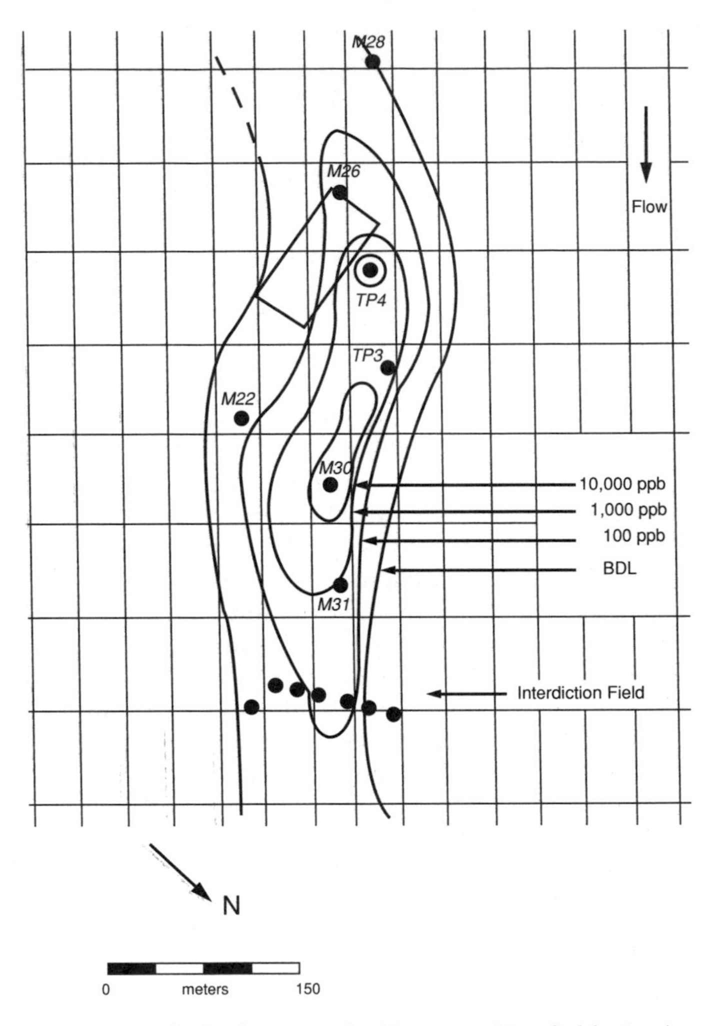

FIGURE 1 Aviation fuel plume at the Traverse City field site (quarter 2, 1986). SOURCE: Rifai et al. (1988). Journal of Environmental Engineering, Vol. 114:1021. Copyright © by the American Society of Civil Engineers (ASCE). Reprinted with permission of ASCE.

and xylene (BTX) were shown to be strongly correlated by statistical analyses and solute transport modeling using BIOPLUME II. The results were remarkably consistent with field data on the presence of high or low levels of BTX and dissolved oxygen in several monitoring well samples.

REFERENCES

Borden, R. C., and P. B. Bedient. 1986. Transport of dissolved hydrocarbons influenced by reaeration and oxygen limited biodegradation: 1. Theoretical development. Water Resources Research 22:1973-1982.

Borden, R. C., P. B. Bedient, M. D. Lee, C. H. Ward, and J. T. Wilson. 1986. Transport of dissolved hydrocarbons influenced by oxygen limited biodegradation: 2. Field application. Water Resources Research 22:1983-1990.

Chiang, C. Y., J. P. Salanitro, E. Y. Chai, J. D. Colthart, and C. L. Klein. 1989. Aerobic biodegradation of benzene, toluene, and xylene in a sandy aquifer—data analysis and computer modeling. Ground Water 27(6):823-834.

Kissel, J. C., P. L. McCarty, and R. L. Street. 1984. Numerical simulation of mixed-culture biofilm. American Society of Civil Engineers Journal of Environmental Engineering Division 110:393.

Konikow, L. F., and J. D. Bredeheoft. 1978. Computer Model of Two-Dimensional Solute Transport and Dispersion in Ground Water: Automated Data Processing and Computations. Techniques of Water Resources Investigations of the U.S. Geological Survey. Washington, D.C.: U.S. Geological Survey.

MacQuarrie, K. T. B., and E. A. Sudicky. 1990. Simulation of biodegradable organic contaminants in groundwater: 2. Plume behavior in uniform and random flow fields. Water Resources Research 26(2):223-239.

MacQuarrie, K. T. B., E. A. Sudicky, and E. O. Frind. 1990. Simulation of biodegradable organic contaminants in groundwater: 1. Numerical formulation in principal directions. Water Resources Research 26(2):207-222.

McCarty, P. L., M. Reinhard, and B. E. Rittmann. 1981. Trace organics in groundwater. Environmental Science and Technology 15(1):40-51.

Molz, F. J., M. A. Widdowson, and L. D. Benefield. 1986. Simulation of microbial growth dynamics coupled to nutrient and oxygen transport in porous media. Water Resources Research 22:107.

National Research Council. 1990. Ground Water Models: Scientific and Regulatory Applications. Washington, D.C.: National Academy Press.

Rifai, H. S., P. B. Bedient, R. C. Borden, and J. F. Haasbeek. 1987. BIOPLUME II Computer Model of Two-Dimensional Contaminant Transport Under the Influence of Oxygen Limited Biodegradation in Ground Water User's Manual Version 1.0. Houston: Rice University, National Center for Ground Water Research.

Rifai, H. S., P. B. Bedient, J. T. Wilson, K. M. Miller, and J. M. Armstrong. 1988. Biodegradation modeling at a jet fuel spill site. American Society of Civil Engineers Journal of Environmental Engineering Division 114:1007-1019.

Srinivasan, P., and J. W. Mercer. 1988. Simulation of biodegradation and sorption processes in ground water. Ground Water 26(4):475-487.

Widdowson, M. A., F. J. Molz, and L. D. Benefield. 1987. Development and application of a model for simulating microbial growth dynamics coupled to nutrient and oxygen transport in porous media. Pp. 28-51 in Proceedings of the Association of Ground Water Scientists and Engineers/International Ground Water Model Center, Holcomb Research Center Institute Conference on Solving Ground Water Problems with Models. Dublin, Ohio: National Ground Water Association.

Widdowson, M. A., F. J. Molz, and L. D. Benefield. 1988. A numerical transport model for oxygen- and nitrate-based respiration linked to substrate and nutrient availability in porous media. Water Resources Research 24(9):1553-1565.

Testing Bioremediation in the Field

John T. Wilson
U.S. Environmental Protection Agency
Robert S. Kerr Environmental Research Laboratory
Ada, Oklahoma

SUMMARY

An operational definition for success of in situ bioremediation at field scale includes meeting regulatory goals for ground water quality in a timely fashion at a predictable cost. Current practice for site characterization does not adequately define the amount of contamination subject to bioremediation. As a result, laboratory estimates of the requirements for electron acceptors and mineral nutrients and of the time required for remediation have much uncertainty. Another aspect of success is the capacity to continue to meet regulatory goals for ground water quality after the active phase of remediation is complete. In contrast to laboratory studies, the extent of remediation achieved at field scale is influenced by dilution of compounds of regulatory concern in circulated water and by partitioning of the regulated compounds between water and residual nonaqueous-phase oily material. The extent of weathering of residual oily-phase material and the hydrologic environment of the residual have a strong influence on the potential for ground water contamination after active remediation ceases.

INTRODUCTION

Transfer of bioremediation laboratory research to the field is often a frustrating and unsatisfying activity. Part of the problem has to do with the levels of inquiry in the laboratory and in the field. Laboratory studies deal with biochemical or physiological processes. Appropriate controls ensure that only one mechanism is responsible for the phenomena under study. During field-scale implementation of bioremediation technology, several processes operate concurrently. They may involve several distinct mechanisms for biological destruction of the contaminant, as well as partitioning to immobile phases, dilution in ground water, and volatilization.

Experimental controls are usually unavailable during full-scale implementation of in situ bioremediation because the technology is applied uniformly to the contaminated area. As a result, performance monitoring that is limited to the concentration of contaminants in ground water over time, and perhaps the concentrations of nutrients and electron acceptors, cannot ensure that the biological process developed in the laboratory was responsible for contaminant removal at full scale.

The appropriate equivalent of experimental controls is a detailed characterization of the site, the flow of remedial fluids, and the flux of amendments. This characterization allows an assessment of the influence of partitioning, dilution, or volatilization and provides a basis for evaluating the relative contribution of bioremediation.

APPROPRIATE SITE CHARACTERIZATION

Most plumes of organic contamination in ground water originate from spills of refined petroleum hydrocarbons, such as gasoline, or chlorinated solvents, such as trichloroethylene. These substances enter the subsurface as nonaqueous-phase oily liquids, traveling separately from the ground water. As long as the oily-phase liquid is present in the subsurface, it can act as a continuing source of contamination because contaminants contained in the nonaqueous phase will dissolve in the ground water.

Traditionally, monitoring wells have been used to define the extent of contamination in the subsurface environment. However, wells cannot determine the extent of contamination by oily-phase materials. If monitoring well data are the only data available, it is difficult to estimate the total contaminant mass subject to remediation within an order of magnitude.

As an example, Kennedy and Hutchins (1992) estimated the mass of alkylbenzenes released to a shallow water table aquifer from a pipeline spill of refined petroleum products. The contaminated area was roughly circular with a diameter of 150 m. Thirteen monitoring stations were located uniformly across the spill. At each station a series of continuous cores was taken extending from clean material above the spill, through the spill, to clean material below it. Then monitoring wells were installed in the boreholes used to acquire the cores.

The cores were extracted and analyzed for the content of BTEX (benzene, toluene, ethylbenzene, xylenes) and total petroleum hydrocarbons. Ground water was collected and analyzed for the same parameters. Data from the 13 stations were subjected to geostatistical analysis to estimate the total contaminant mass in the aquifer and the total mass dissolved in the ground water (see Clark, 1979, for a description of geostatistics). Kennedy and Hutchins (1992) used proprietary computer software to estimate the total mass of contaminants. SURFER, available from Golden Software, Denver, Colorado, supports linear kriging of plan two-dimensional data following the trapezoidal rule, Simpson's rule, and Simpson's 3/8 rule. Lacking information on the structure of the data, Kennedy and Hutchins ran all three simulations and took the numerical average.

Of 320 kg of benzene in the aquifer, only 22 kg was dissolved in the ground water; of 8800 kg of BTEX compounds, only 82 kg was dissolved; and of 390,000 kg of total petroleum hydrocarbons, only 115 kg was dissolved. Monitoring well data grossly underestimated the extent of contamination. This research shows the need for site characterization techniques that can accurately estimate the total mass of contaminants subject to bioremediation.

Estimates of Total Contaminant Mass

Estimates of total contaminant mass in the subsurface are required to predict the demand for nutrients and electron acceptor that must be met to complete the remediation. If the total demand for nutrients and electron acceptor has been estimated, the rate of supply of the limiting requirement can be used to estimate the time required for remediation.

The most rigorous approach involves the collection of cores from the contaminated regions of the subsurface environment, followed by extraction and analysis of the cores for the contaminants of concern. A wireline piston sampler (Zapico et al., 1987) makes it possible to collect representative continuous cores, even in noncohesive material.

To preclude losses to volatilization and biodegradation, current good practice recommends against shipping core samples to a laboratory for extraction. When cores are subsampled and extracted in the field, the variability between replicate subsamples is much smaller (Siegrist and Jenssen, 1990). Standard operating procedure at the Robert S. Kerr Environmental Research Laboratory uses a paste sampler to take replicate subcores of each core sample. Each subcore weighs 10 to 15 g and represents 10 to 15 cm of vertical core. The subcore is delivered to a 40-ml vial, containing 5 ml of methylene chloride, which is sealed with a Teflon-faced septum. The sample is dispersed in the solvent to begin extraction, preclude volatilization, and prevent biodegradation. The vials are shipped to the laboratory for subsequent extraction and analysis.

Unless the cores are screened to identify those that deserve analysis, this approach may be too expensive to be practical for general use. The data set reported by Kennedy and Hutchins (1992) contained more than 400 analyses. However, there are several techniques that can reduce the analytical burden. Headspace analysis can be used to screen cores in the field to determine whether the depth interval represented by a core is contaminated with oily-phase hydrocarbons, so that field extraction and analysis of the core are justified.

Aquifer samples can be equilibrated with the headspace of a plastic bag before analysis by a field gas chromatograph, organic vapor analyzer, or explosimeter (Robbins et al., 1989). Alternately, a plug can be removed from a core with a paste sampler and the inlet to an explosimeter or organic vapor analyzer inserted directly into the cavity (Kampbell and Cook, 1992). These techniques are inexpensive and generate data in real time, which allows the screening information to be used to guide decisions about depth and location of subsequent cores.

Often the meter's response in the field headspace analyses has a strong correlation to the content of total petroleum hydrocarbons. In these cases, results from a limited number of expensive core analyses can be extrapolated to a large number of inexpensive field headspace analyses. Kampbell and Cook (1992) compared the hydrocarbon vapor concentrations in the cavities left when plugs were removed from cores for extraction to the content of total petroleum hydrocarbons in the cores. The correlation coefficient between meter response and hydrocarbon content was 0.957 on a set of 24 cores and 0.801 on a set of 64 cores.

Estimates of Contaminant Dilution in Ground Water

In the process of remediation, prodigious quantities of water may be circulated through an oily-phase spill. The compounds of regulatory concern, such as the BTEX compounds, are often more water soluble than the other components of refined petroleum hydrocarbons. As the circulated water sweeps through the spill, the more water-soluble components partition into the water and are diluted out. The concentration of regulated compounds will drop because of simple dilution.

Downs et al. (1989, in press) quantitatively described this effect in a pilot-scale demonstration of in situ bioremediation of a jet fuel spill using nitrate as the electron acceptor. They used a tracer to estimate the volume of recirculated water, and they cored the area perfused by ground water to estimate the quantity of total hydrocarbons and the quantity of hydrocarbons of regulatory concern. Simple partitioning theory was used to calculate the distribution of hydrocarbons of concern between recirculating ground water and the residual jet fuel.

Estimate of Recirculated Volume of Water

An infiltration gallery sited above the spill effectively perfused a plan surface area of 130 m^2 with nitrate-amended ground water. The infiltrated water was recovered in five purge wells. In Figure 1 a computer model predicts flow paths from the infiltration gallery to the recovery wells. In Figure 2 a cross section shows the relationship of the infiltration gallery, the contaminated interval, and the recovery wells. A pulse of chloride was used to trace the flow of water to the recovery wells. The volume of circulated water was considered to be the pumping rate multiplied by the travel time of the chloride between infiltration and the recovery wells. The arrival time at each well was weighted by its pumping rate to calculate the overall residence time, and circulated volume, between infiltration and recovery. Figure 3 shows breakthrough curves of the chloride tracer.

In the demonstration the average residence time was 10 days and the circulated volume was 10,900 m^3.

Estimate of Partitioning Between Oil and Water

The area of the spill that was perfused by the infiltration gallery was cored to determine the total quantity of jet fuel and the quantities of benzene, toluene, ethylbenzene, and the xylenes. Smith et al.

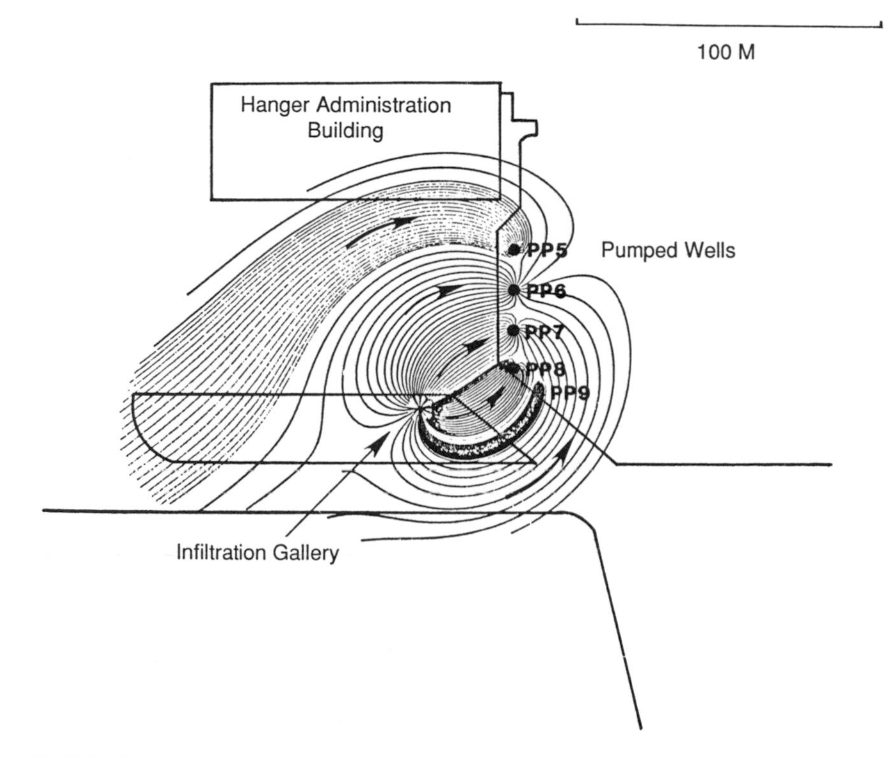

FIGURE 1 Hydraulic model of in situ bioremediation of a JP-4 spill at the Traverse City demonstration site. Ground water amended with mineral nutrients and nitrate as an electron acceptor was recharged through an infiltration gallery. The model predicted flow lines from the infiltration gallery to pumping wells (PP5 to PP9) that capture and recirculate ground water. To capture the infiltrated water containing nitrate, more water was pumped than was delivered to the gallery. As a result, some of the flow lines to the wells originate in uncontaminated ground water upgradient of the spill.

(1981) reported empirical partition coefficients for the compounds between JP-4 jet fuel and water. To estimate the distribution of an individual BTEX compound between fuel and water, the published partition coefficients were multiplied by the ratio of the volume of JP-4 under the infiltration gallery to the volume of water in circulation. The distribution between oil and fuel was used to calculate the fraction of total material in oil or water. To predict the equilibrium solution concentration of a BTEX compound in circulation, the quantity of the compound originally in the fuel was multiplied by the fraction that should partition to water and was divided by the circulated volume of ground water.

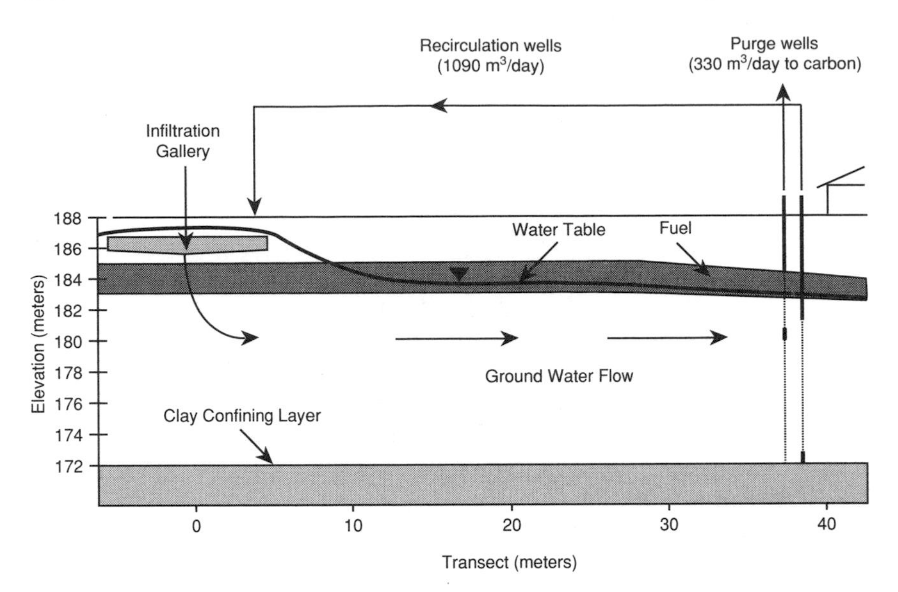

FIGURE 2 Cross section of a JP-4 spill at the Traverse City demonstration site. Ground water was amended with mineral nutrients and nitrate as an electron acceptor. Water was recirculated to an infiltration gallery installed above the JP-4 spill. It moved vertically across the spill, then laterally through the aquifer to the recovery wells. Part of the recovered water was recirculated; part was purged.

Table 1 compares the predicted dilution of benzene, toluene, and o-xylene to the actual concentration in monitoring wells before recirculation of ground water. Dilution alone produced at least a fivefold reduction in concentration.

To maintain hydraulic control over the spill, a fraction of the recovered water was discharged to waste. This flow was replaced with clean water from the aquifer. If the circulation system behaved as a completely mixed reactor, solutes in the circulated water would be diluted at a first-order rate of 0.03 per day. Dilution of a BTEX compound was estimated by multiplying the rate of dilution of circulated water by the portion of the total mass that partitioned to the circulated water.

Estimate of Bioremediation

The actual behavior of benzene is depicted in Figure 4. The dashed line is the calculated equilibrium concentration of benzene in the recirculated water based on partitioning and dilution. The solid line

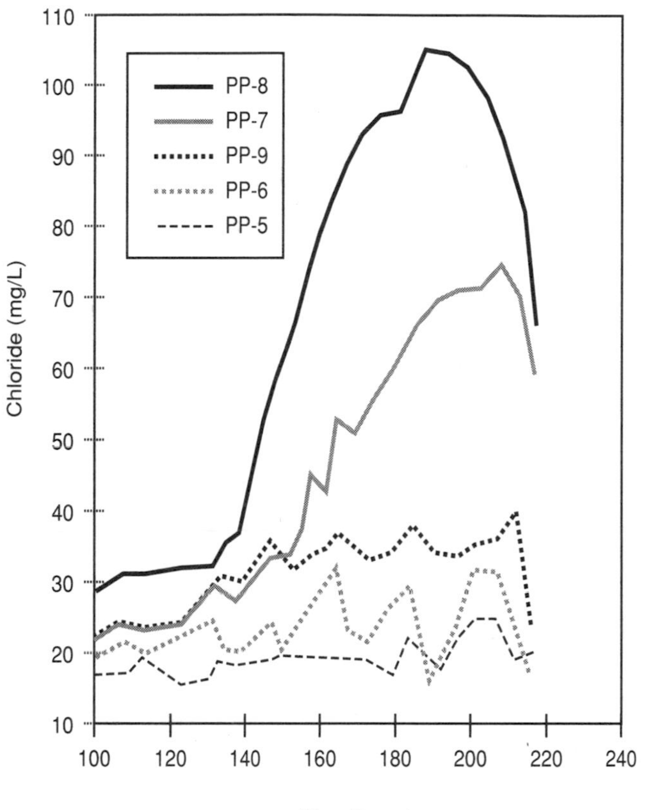

FIGURE 3 Breakthrough of chloride from a tracer test conducted to evaluate the hydraulic model of water flow. Chloride was added to water supplied to the infiltration gallery. The concentration of chloride that breaks through is proportional to the number of flow lines originating in the infiltration gallery, compared to flow lines originating upgradient in the aquifer.

shows concentrations in the recirculation well that captured the greatest portion of infiltrated water. Concentrations rose slowly over time, overshot the prediction at about two recirculation volumes, then showed good agreement with the prediction for another recirculation volume. Then biological acclimation occurred, and benzene was removed from the circulated water. Concentrations dropped below the analytical detection limit over a 2-day period. Concentrations of other BTEX compounds were not reduced (compare data for o-xylene in Figure 5), which established that the removal was a biological process. If a physical or chemical process had been responsible for the

TABLE 1 Comparison of Concentrations (μg/l) of BTEX Compounds in Ground Water After Bioremediation to the Concentrations Expected from the BTEX Content of the Residual Petroleum Hydrocarbons

Compound	Concentration Prior to Bioremediation	Concentration After Bioremediation	Concentration Predicted from Residual
Benzene	760	<1	2
Toluene	4500	<1	15
Ethylbenzene	840	6	6
m,p-Xylene	2600	23	27
o-Xylene	1380	37	18

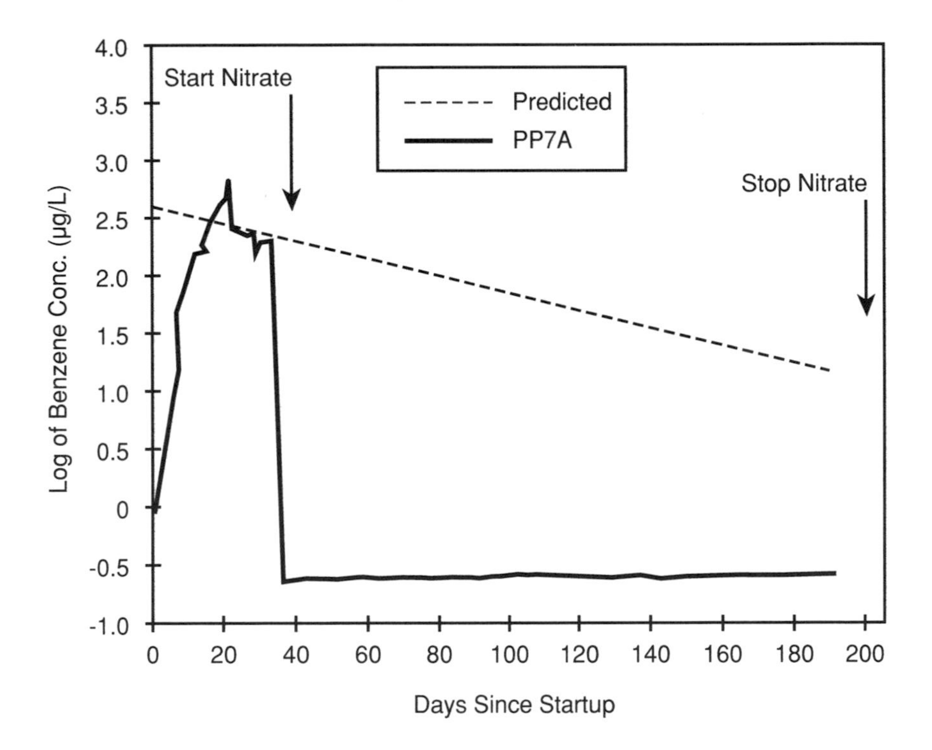

FIGURE 4 Bioremediation of a JP-4 jet fuel spill using nitrate. Comparison of the depletion of benzene in circulated ground water to the depletion predicted from dilution and partitioning.

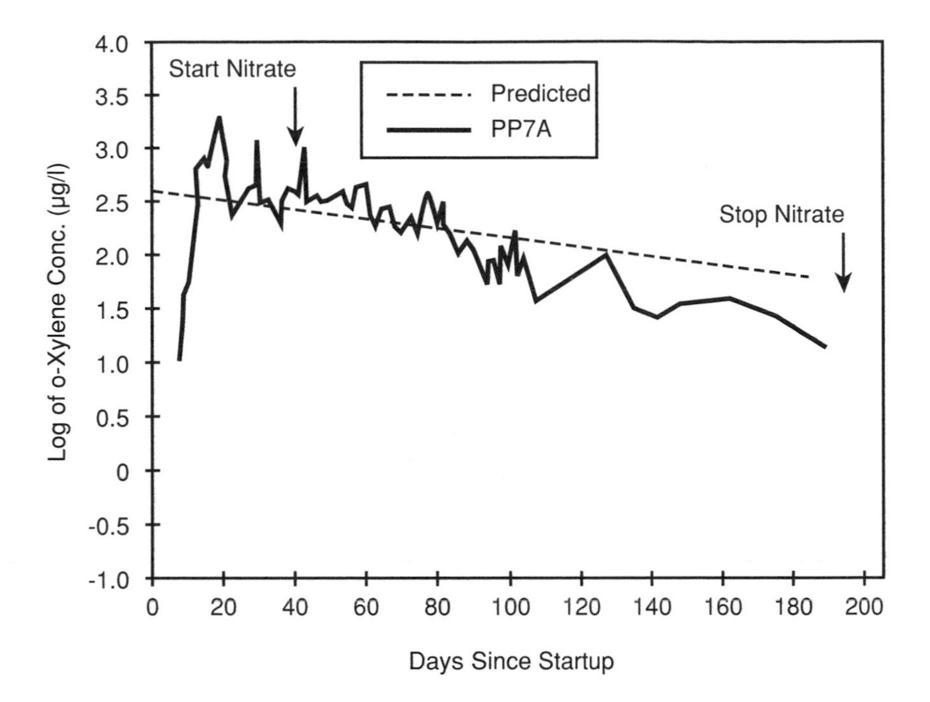

FIGURE 5 Comparison of the depletion of o-xylene in circulated ground water to the depletion predicted from dilution and partitioning in a JP-4 jet fuel spill.

benzene removal, other BTEX compounds (such as o-xylene) would also have been removed because these other compounds have physical and chemical properties similar to those of benzene but are less readily biodegradable.

Note from Figure 4 that removal of benzene occurred before nitrate was added to the system. This effect had not been predicted in the laboratory treatability study conducted as part of the design of the pilot-scale demonstration (Hutchins et al., 1991a) and has not been conclusively reproduced at laboratory scale since then.

CRITERIA FOR SUCCESS AT FIELD SCALE

The criteria for success at each remediation site are unique, depending on the particular requirements of state and federal regulators and the particular concerns of the site owner. However, requirements at many sites can be generalized to the following:

1. Concentrations of substances of regulatory concern in ground water at the end of active bioremediation will be less than the cleanup goals established by the regulatory authorities.

2. Concentrations of substances of regulatory concern in ground water will not rise above the cleanup goals, within a prescribed period of monitoring, after active bioremediation is concluded.

3. The site owner will enjoy beneficial use of the property during remediation and will be allowed to sell or transfer it when remediation has been complete.

The following considerations have a direct bearing on the first two requirements.

Can Any Oily-Phase Residual Support a Plume?

Monitoring wells can provide a misleading picture of the course of bioremediation. Pumping, or seasonal changes in regional water tables, can drop ground water elevations below the depth interval occupied by oily-phase contaminants. Water produced by monitoring wells may be clean, but contamination will return when pumping stops or recharge raises the regional water table elevation. Changes in the stage of nearby rivers or lakes, combined with seasonal variations in recharge, may alter the slope of the water table (hydraulic gradient), which will change the trajectory of the plume of contamination. Plumes may actually move away from monitoring wells under these conditions, then return to them later.

To supplement data from monitoring wells, many regulatory authorities require a measure of residual oily-phase material left after bioremediation. Cleanup goals are usually set with the conservative assumption that the relative composition of oily-phase material does not change during remediation. As a result, concentrations of oily-phase material that are determined to be protective of ground water quality are low, on the order of 10 to 100 mg total petroleum hydrocarbon per kilogram of aquifer material (Bell, 1990).

Bioremediation, particularly innovative bioremediation that uses an electron acceptor other than oxygen, can remove the compounds of regulatory concern from the subsurface while leaving significant amounts of oily-phase hydrocarbons. The issue is whether any residual oily-phase hydrocarbon is capable of producing a plume of contamination at concentrations that exceed the cleanup goal.

The JP-4 bioremediation demonstration at Traverse City, Michigan, was used to evaluate the importance of partitioning of contaminants between ground water and residual oily material. The concentrations of BTEX compounds in recirculated ground water were compared to concentrations in the weathered oily-phase residual.

When infiltration with nitrate brought the concentrations of BTEX compounds in the JP-4 spill below action levels, infiltration was stopped and concentrations of BTEX compounds in the aquifer were measured under natural conditions. The JP-4 contaminated interval was cored and analyzed for residual total hydrocarbons and concentrations of BTEX compounds. The reduction in concentration of total petroleum hydrocarbons was minimal, from 2000 mg/kg to 1400 mg/kg. The concentrations of BTEX compounds in the residual oil were divided by the fuel-to-water partition coefficients of Smith et al. (1981) to predict the capacity of the residual to contaminate ground water. The concentrations measured in ground water under natural conditions were near or below the predicted concentrations (Hutchins et al., 1991b; see Table 1).

Apparently ground water quality is controlled by the relative concentration of organic contaminants in the weathered oily-phase residual and not by the absolute amount of weathered total petroleum hydrocarbons. The relative concentrations of organic contaminants can be used to predict the concentrations in ground water in contact with the oily-phase residual.

Accounting for Spatial Heterogeneity

Bioremediation is difficult to assess in heterogeneous geological material. Often, oily-phase material is associated with fine-textured material with low hydraulic conductivity. Remedial fluids tend to pass around the fine-textured material. Because the flux of nutrients and electron acceptor through the fine-textured material is small, there is little opportunity for bioremediation, and significant concentrations of contaminants can remain in subsurface material.

These relationships will be illustrated in a case history from an industrial site in Denver, Colorado (Nelson et al., in press). At this site, a temporary holding tank under a garage leaked used crankcase oil, diesel fuel, gasoline, and other materials into a shallow water table aquifer. Figure 6 shows the relationship between the garage, the work pit containing the leaking holding tank, and the approximate area of the spill.

Remediation involved removal of separate oily phases, in situ bioremediation with hydrogen peroxide and mineral nutrients, and bioventing. Ground water flow under ambient conditions was to the north or northeast. The flow of water during the remediation paralleled the natural gradient. Water was produced from a recovery well

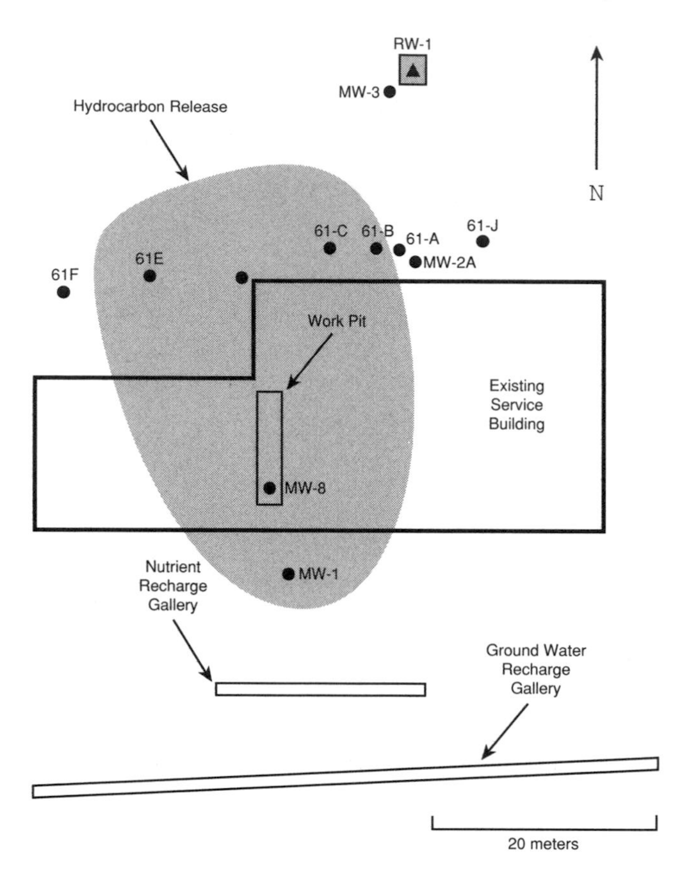

FIGURE 6 Infrastructure at an in situ bioremediation project in Denver, Colorado. A holding tank in a work pit under a garage leaked petroleum hydrocarbons to the water table aquifer. Ground water was pumped from a recovery well (RW-1) and filtered through activated carbon. The flow was split. Part was amended with hydrogen peroxide and mineral nutrients and recharged in a nutrient recharge gallery. The remainder was recharged in a ground water recharge gallery. The system was designed to sweep hydrogen peroxide and nutrients under the service building. MW-1, 2A, and 3 are monitoring wells, and 61A through 61J are boreholes for cores.

on the northeast side of the spill. The flow from the well was split; part of the flow was amended with hydrogen peroxide and nutrients and recharged to the aquifer in a nutrient recharge gallery on the south side of the spill (Figure 6). The remainder of the flow was delivered to a ground water recharge gallery to the south of the nutrient recharge gallery. From 3 to 6 gpm (11 to 22 l/min) was deliv-

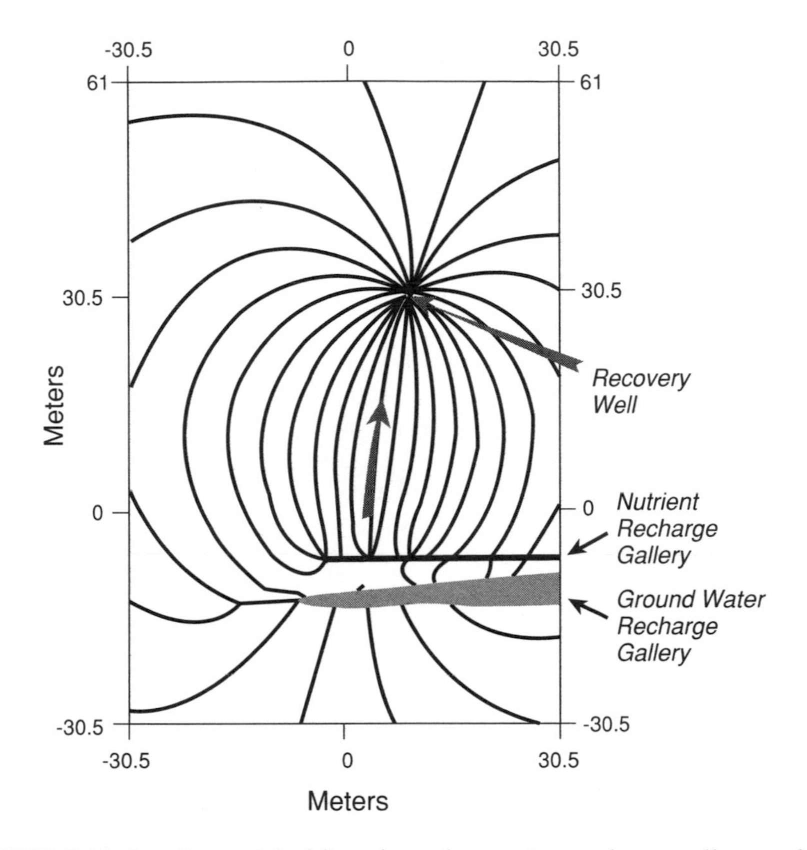

FIGURE 7 Hydraulic model of flow from the nutrient recharge gallery to the recovery well. Note that the spill is contained within the flow path from the gallery to the well and that the well captures all the flow lines from the nutrient recharge gallery.

ered to the nutrient recharge gallery, and 4 to 8 gpm (15 to 30 l/min) was delivered to the ground water recharge gallery, for a total flow of 9 to 11 gpm (34 to 42 l/min). Figure 7 presents a mathematical model of the flow paths from the galleries to the recovery well. The system was designed to sweep the ground water containing hydrogen peroxide and mineral nutrients through the spill to the recovery well.

The system was operated from October 1989 to March 1992. At a flow of 10 gpm (38 l/min), 10 to 15 pore volumes would have been exchanged in the area between the nutrient recharge gallery and the recovery well.

Remediation of Ground Water at the Denver Site

Table 2 compares the reduction in concentrations of benzene and total BTEX compounds in ground water that was achieved by in situ bioremediation at the site in Denver. Monitoring wells MW-1 and MW-8 (shown in Figure 6) are in areas with oily-phase hydrocarbons. Well MW-2A is just outside the region with oily-phase hydrocarbon. Well MW-3 is a significant distance from the region with oily-phase hydrocarbon; it sampled the plume of contaminated ground water that moved away from the spill.

Before remediation, concentrations in wells MW-1 and MW-8 were equivalent. Well MW-1 was closest to the nutrient recharge gallery, and the aquifer surrounding MW-1 was completely remediated; BTEX compounds were undetectable in ground water. In well MW-8, immediately adjacent to the point of release, the concentration of benzene was reduced at least one order of magnitude, and the concentrations of benzene and BTEX compounds in well MW-3 were also reduced an order of magnitude.

It is of particular interest that significant concentrations of benzene or total BTEX never developed in the pumped recovery well (RW-1). The BTEX compounds were monitored twice a month from July 1989 to March 1992. Benzene was detected only twice, at a concentration of 2 μg/l. The other BTEX compounds were never detected. Water from contaminated flow paths sampled by MW-3 was probably diluted by uncontaminated water from other flow paths to RW-1 (compare Figure 7). This behavior illustrates the contrast in contaminant concentrations between passive monitoring wells and pumped wells.

TABLE 2 Reduction in Concentration (μg/l) of Hydrocarbon Contaminants in Ground Water Achieved by In Situ Bioremediation

	Benzene			Total BTEX		
Well	Before	During	After	Before	During	After
MW-1	220	<1	<1	2030	164	<6
MW-8	180	130	16	1800	331	34
MW-2A	?	11	0.8	?	1200	13
MW-3	11	5	2	1200	820	46
RW-1	<1	2	<1	<1	2	<1

Remediation of Subsurface Material at the Denver Site

Water in the monitoring wells and the recirculation well contained low concentrations of contaminants by March 1992. Active remediation was terminated, and the site entered a period of postremediation monitoring.

In June 1992 core samples were taken from the aquifer to determine the extent of hydrocarbon contamination remaining and whether a plume of contamination could return once active remediation ceased. The site was cored along a transect downgradient of the release. The transect extended laterally from clean material, through part of the spill, into clean material on the other side. In each borehole, continuous cores extended vertically from clean material above the spill, through the spill, to clean material below. The cores were extracted and analyzed for total petroleum hydrocarbons and for the concentrations of individual BTEX compounds.

The relationships between the land surface, the water table, the region containing hydrocarbons, and the bedrock are presented in Figure 8. Significant amounts of hydrocarbons remain within a narrow interval, approximately 0.6 m thick, near the water table. The total saturated thickness of the aquifer was approximately 6 m. At the time of sampling the elevation of the water table was 1610 m (5280.5 ft) above mean sea level, and all the hydrocarbons were below the water table.

The highest concentrations of hydrocarbons at the Denver site were obtained in samples from the borehole (D) closest to the work pit. Table 3 presents the vertical distribution of BTEX compounds and total petroleum hydrocarbons (TPH) in borehole D. The material in the interior of the spill had higher proportions of BTEX compounds. Table 4 makes the same comparison at the most contaminated depth interval along the transect. Material closer to the spill had higher concentrations of TPH and greater relative proportions of BTEX compounds.

Figure 9 plots the percentage of BTEX in the residual oil after bioremediation against the total content of hydrocarbon. Obviously, the materials with lower residual concentrations of hydrocarbons are more extensively weathered.

Infiltration of hydrogen peroxide and mineral nutrients at an aviation gasoline spill in Michigan preferentially removed BTEX compounds from the oily-phase gasoline, leaving a total petroleum hydrocarbon residual low in aromatic hydrocarbons (Wilson et al., in press). At the Denver site, apparently, a cortex of material that has been physi-

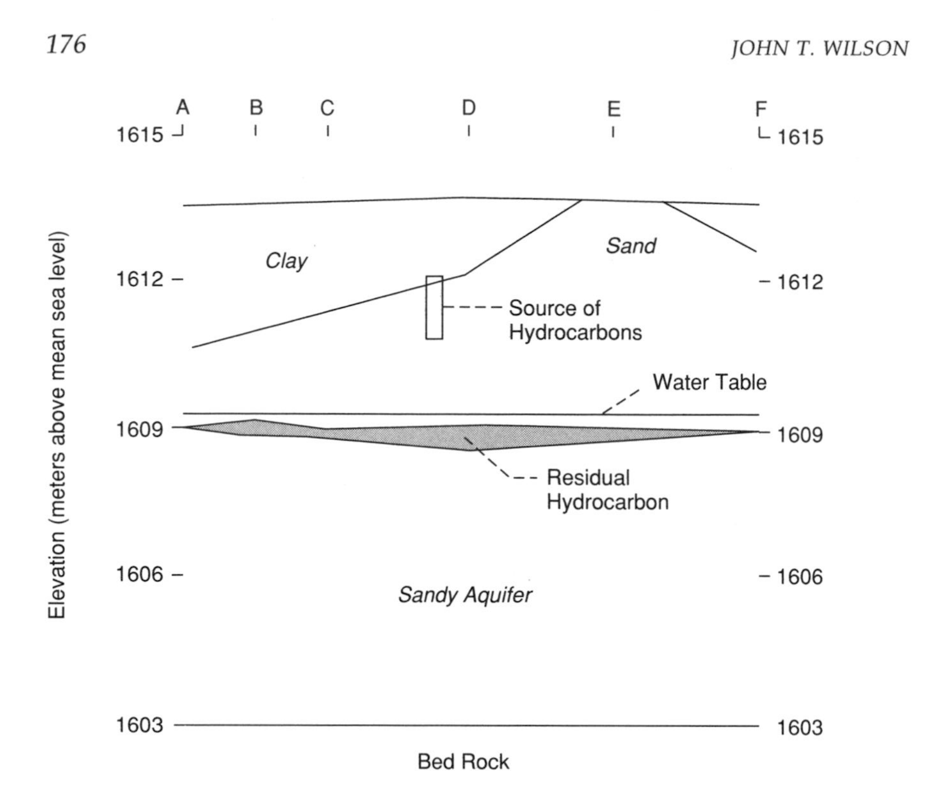

FIGURE 8 Cross section showing the vertical relationship of the land surface, water table, residual hydrocarbon after bioremediation, and the lower confining layer of the aquifer. The cross section runs through core boreholes depicted in Figure 6.

TABLE 3 Vertical Extent of Total BTEX Compounds and Total Petroleum Hydrocarbons (mg/kg) at Borehole D, the Most Contaminated Borehole in the Transect (Figure 8)

Elevation[a]	TPH	BTEX	Benzene	Color and Texture
1609.711 to 1609.458	<44	<1	<0.2	Brown sand
1609.458 to 1609.354	227	5.1	<0.2	Brown sand
1609.354 to 1609.230	860	101	<0.2	Black sand
1609.230 to 1609.101	1176	206	4.3	Black sand
1609.101 to 1609.050	294	27	0.68	Black sand
1609.050 to 1608.949	273	7.4	0.26	Black sand
1608.949 to 1608.821	<34	<1	<0.2	Black sand
1608.821 to 1608.492	<24	<1	<0.2	Brown to yellow sand

[a]Meters above sea level.

TABLE 4 Lateral Distribution of Total BTEX Compounds and Total Petroleum Hydrocarbons (mg/kg) Along the Transect (Figure 8) at the Most Contaminated Depth Interval

Borehole	TPH	BTEX	Benzene
B	167	0.8	<0.2
C	156	3.5	<0.2
D	1176	260	4.3
E	156	3.5	0.06

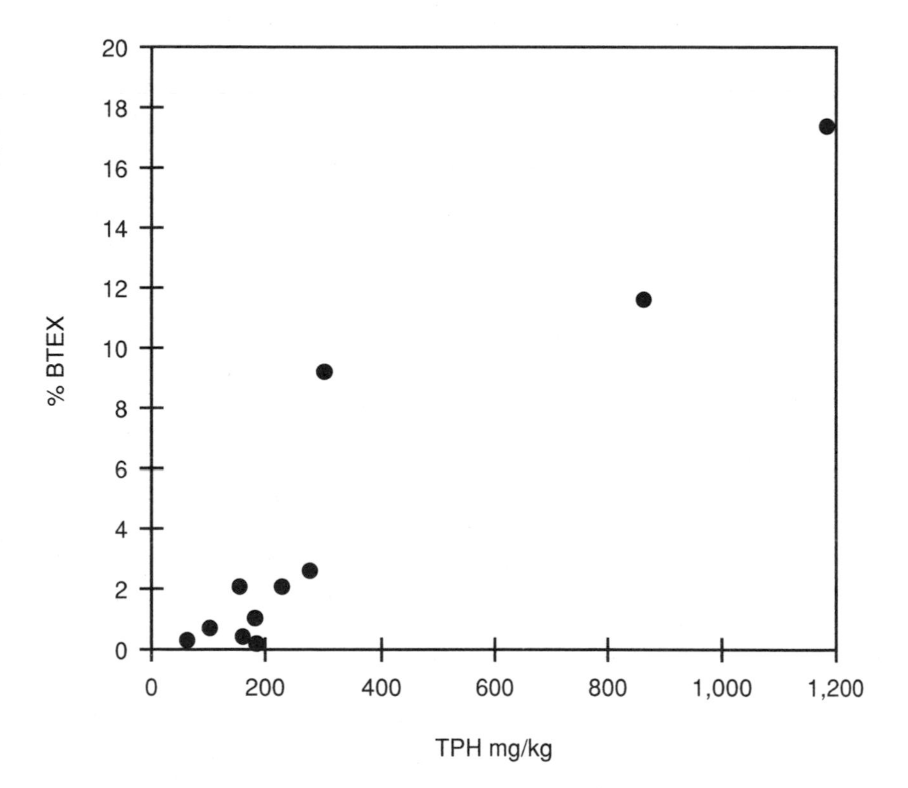

FIGURE 9 Relationship between extent of hydrocarbon contamination (concentration of total petroleum hydrocarbons) and the extent of biological and chemical weathering (reduction in percentage of BTEX compounds in total petroleum hydrocarbons) in core material after bioremediation at the Denver site.

cally and biologically weathered surrounds a central core of material that has not been depleted of BTEX compounds.

The concentration of an individual petroleum hydrocarbon in solution in ground water in contact with oily-phase hydrocarbon can be predicted by Raoult's law. The solution concentration in water should be proportional to the mole fraction of the hydrocarbon in the oily phase. Assume that the weathered material is weathered because it is in effective contact with moving ground water that supplied nutrients and electron acceptors, and the residual is not weathered because it was not in effective contact and the supply of nutrients and electron acceptors was inadequate. If partitioning between moving ground water and the weathered oily residual controls the concentration of hydrocarbons in the water, the 10-fold reduction in concentrations of benzene and BTEX compounds seen in the weathered core material (Table 3) would produce the 10-fold reduction in concentrations of benzene and BTEX compounds seen in the monitoring wells (Table 2).

Do Mass Transfer Effects Limit Development of a Plume?

The usual expectation for in situ bioremediation is total removal of the contaminant from the subsurface environment. The extent of remediation more commonly achieved is removal of the contaminant from the circulated ground water.

In situ bioremediation merely accelerates the natural physical and biological weathering processes that occur in the subsurface. The oily material in most intimate contact with the circulated ground water is weathered to the greatest extent. After extensive remediation of the more transmissive regions, the release of contaminants to the circulated ground water is controlled by diffusion and slow advection from the subsurface material that still contains significant quantities of contaminants. The relationships are presented in Figure 10.

In such circumstances the disposition of contamination in the ground water can only be understood as a dynamic system. This release may best be described through the chemical engineering concept of a mass transfer coefficient. As the circulated water passes through the weathered spill, a certain quantity of hydrocarbon is transferred to the water; the amount transferred is directly proportional to the exposure time of the water in the contaminated area. If the circulated water contains enough nutrients and electron acceptor to meet the demand of the contaminants transferred from the fine-textured material, the plume will be destroyed by biological activity as rapidly as it is produced.

When active remediation is stopped, the concentration of elec-

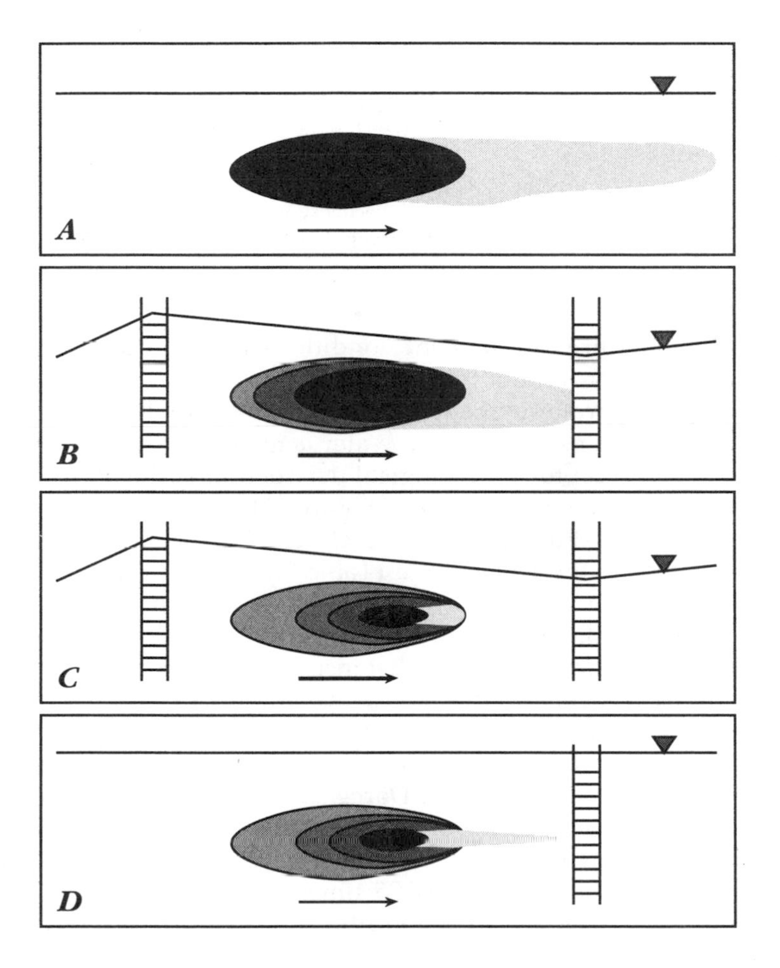

FIGURE 10 Schematic representation of in situ bioremediation in heterogeneous geological material. A fresh release (panel A) is rapidly weathered (panel B) due to increased flow of water and increased concentration of electron acceptor. As weathering progresses, aromatic hydrocarbons such as the BTEX compounds are restricted to regions with low hydraulic conductivity (panel C). After bioremediation the flux of aromatic hydrocarbons from the residual core to the moving ground water is controlled by mass transport limitations. The extent of the plume produced is controlled by the supply of electron acceptor (panel D). Although greatly attenuated, the plume may be regenerated under ambient conditions.

TABLE 5 Contrast in Conditions During Active In Situ Bioremediation and Conditions at the Termination of Remediation at a Spill from an Underground Storage Tank in Denver Colorado

Parameter	During Active Remediation	Ambient Conditions After Remediation
Introduced concentration of oxygen	470 mg/l	5.5 mg/l
Hydraulic gradient	0.097 m/m	0.0012 m/m
Interstitial flow velocity	2.4 m/day	0.03 m/day
Travel time of water across the spill	20 days to recovery well	1500 days to monitoring well
Maximum oxygen demand supported	20 mg/l per day	0.004 mg/l per day

tron acceptor returns to ambient conditions in the aquifer, and the hydraulic gradient returns to the normal condition. As a result, the residence time of water in the spill area is longer, and the total amount of hydrocarbon transferred to the water is greater, although the supply of electron acceptor for biological destruction of the hydrocarbon is less (compare panels C and D in Figure 10).

These relationships are well illustrated by the performance of bioremediation at the Denver site (Table 5). The hydraulic conductivity in the depth interval containing the hydrocarbons is approximately 8.5 m per day (David Szlag, University of Colorado, Boulder, personal communication). The distance from the nutrient recharge gallery to the recovery well is 45 m. Assume that the distance from the upgradient edge of the spill, through the spill, to monitoring wells downgradient at the property boundary is also 45 m. Assuming an effective porosity of 0.35, Darcy's law can be used to estimate the interstitial flow velocity of the ground water and its residence time along the flow path. When active remediation was terminated, the residence time of water was 75 times longer, and the supply of oxygen was 85 times less than conditions during active remediation (compare Table 5).

In June 1992 core material from the spill was assayed for the potential rate of oxygen consumption. Core material was dewatered by placing it in a Buchner funnel and applying a partial vacuum. Then the core material was sealed in a glass mason jar. After 24 hours, a sample of air in contact with the core material was passed through an oxygen-indicating tube to estimate oxygen consumption potential. Acclimated material exerted oxygen demands of 6 to more than 36 mg/kg core material per day (Table 6), equivalent to 40 to more than 240 mg/l ground water per day. At these rates, the oxy-

TABLE 6 Relationship Between the Potential Oxygen Uptake Rate (mg oxygen per kg per day) of Freshly Collected Core Material and the Location of the Core Material with Respect to the Interval Contaminated with Hydrocarbons

	Position of Borehole in the Transect					
Location in Borehole	A	B	C	D	E	F
Just Above			<4		7.4	
Within	15.5	>30	>36		>34	23.5
Just Below	6.0	<3	5.7	7.3		21

gen supplied as hydrogen peroxide during active remediation would be consumed in 2 to 12 days.

In the assay the microbial consumption of oxygen was faster than the rate of supply during active remediation. If the microbes in the aquifer expressed the potential rate of oxygen consumption, oxygen would have been depleted before the recharge water moved across the spill. In the absence of oxygen, BTEX compounds would have partitioned to the ground water and should have been detected in the monitoring wells. In fact, oxygen concentrations between 2 and 5 mg/l were always present in water produced by the recovery well, and BTEX compounds were virtually absent. Oxygen consumption must have been limited by mass transfer of hydrocarbon to the ground water circulated through the spill.

Relationship to Siting and Sampling Monitoring Wells

No established procedures exist for determining under ambient conditions whether the mass transfer of hydrocarbons from oily residual material will exceed the supply of oxygen or other natural electron acceptors. As a result, it is impossible to predict if natural bioremediation will prevent the regeneration of a plume, or if a plume of contaminated ground water will regenerate and at what concentrations. Ground water moving under the natural gradient must be allowed to travel all the way through the spill and then to the monitoring wells before it is possible to determine whether mass transfer effects will reestablish a plume.

An assessment of natural hydrologic conditions at a site will be necessary to intelligently locate compliance monitoring wells and determine an appropriate schedule of monitoring. Required are an understanding of the average natural hydraulic gradient and the hydraulic conductivity in the depth interval containing residual hydrocarbon. This information can be used to predict the velocity and trajectory of potential plumes of contaminated water. The frequency of monitoring can be adjusted to reflect the expected time required for ground water to travel through the area containing residual hydrocarbon to the point of compliance.

CONCLUSIONS

- When oily-phase materials are to be remediated, core analyses are required to estimate the total mass of contaminant subject to remediation.
- Headspace analyses in the field can be used to screen core samples to identify those that deserve further analysis in the laboratory. If properly benchmarked by a limited number of laboratory analyses, field headspace techniques can provide a rapid and affordable estimate of total contaminant concentrations.
- Simple ground water flow models can estimate the volume of water circulated through a spill during in situ bioremediation. This information can be coupled with simple partitioning theory to estimate the apparent attenuation due to dilution.
- In situ bioremediation frequently leaves a residual of weathered oily-phase material.
- Partitioning theory can be used to predict the concentrations of BTEX compounds in ground water in contact with the weathered oily residual.
- After extensive in situ bioremediation, pockets of fine-textured material may still contain high concentrations of contaminants.
- Mass transfer effects control the access of residual organic contaminants to moving ground water.
- Under proper conditions, natural biodegradation supported by ambient concentrations of electron acceptors and mineral nutrients may destroy organic contaminants as fast as they escape from the oily-phase residual.
- At the present state of science, only long-term monitoring can determine if natural biodegradation will prevent the regeneration of a plume of contaminated ground water.

ACKNOWLEDGMENTS

This work was supported by the United States Air Force through Interagency Agreement RW 57935114 between the Armstrong Laboratory Environics Directorate (U.S. Air Force) and the R. S. Kerr Laboratory (U.S. Environmental Protection Agency). This work was also supported by the U.S. Environmental Protection Agency through the Bioremediation Field Initiative. It has not been subjected to agency review and therefore does not necessarily reflect the views of the agency, and no official endorsement should be inferred.

REFERENCES

Bell, C. E. 1990. State-by-state summary of cleanup standards. Soils: Analysis, Monitoring, Remediation (Nov.-Dec.):10-16.

Clark, I. 1979. Practical Geostatistics. London: Applied Science Publishers.

Downs, W. C., S. R. Hutchins, J. T. Wilson, R. H. Douglas, and D. J. Hendrix. 1989. Pilot project on biorestoration of fuel-contaminated aquifer using nitrate: part I—field design and ground water modeling. Pp. 219-233 in Proceedings, Petroleum Hydrocarbons and Organic Chemicals in Ground Water: Prevention, Detection, and Restoration, Nov. 15-17, 1989, Houston. Dublin, Ohio: National Water Well Association.

Downs, W. C., S. R. Hutchins, J. T. Wilson, R. H. Douglas, and D. J. Hendrix. In press. Nitrate-mediated biodegradation of BTEX in JP-4 contaminated soil and ground water. In Bioremediation: Field Experiences, P. E. Flathman, D. E. Jerger, and J. H. Exner, eds. Chelsea, Mich.: Lewis Publishers.

Hutchins, S. R., G. W. Sewell, D. A. Kovacs, and G. A. Smith. 1991a. Biodegradation of aromatic hydrocarbons by aquifer microorganisms under denitrifying conditions. Environmental Science and Technology 25(1):68-76.

Hutchins, S. R., W. C. Downs, J. T. Wilson, G. B. Smith, D. A. Kovacs, D. D. Fine, R. H. Douglass, and D. J. Hendrix. 1991b. Effect of nitrate addition on biorestoration of fuel-contaminated aquifer: field demonstration. Ground Water 29(4):571-580.

Kampbell, D. H., and M. L. Cook. 1992. Core assay method for fuel contamination during drilling operations. Pp. 139-140 in Subsurface Restoration Conference Proceedings. Houston: Rice University, National Center for Ground Water Research.

Kennedy, L. G., and S. R. Hutchins. 1992. Applied geologic, microbiological, and engineering constraints of in-situ BTEX bioremediation. Remediation 3(1):83-108.

Nelson, C., R. J. Hicks, and S. D. Andrews. In press. In-situ bioremediation: an integrated system approach. In Bioremediation: Field Experiences, P. E. Flathman, D. E. Jerger, and J. H. Exner, eds. Chelsea, Mich.: Lewis Publishers.

Robbins, G. A., R. D. Bristol, and V. D. Roe. 1989. A field screening method for gasoline contamination using a polyethylene bag sampling system. Ground Water Monitoring Review (Fall):87-97.

Siegrist, R. L., and P. D. Jenssen. 1990. Evaluation of Sampling Method Effects on Volatile Organic Compound Measurements in Contaminated Soils. Environmental Science and Technology 24(9):1387-1392.

Smith, J. H., J. C. Harper, and H. Jaber. 1981. Analysis and Environmental Fate of Air Force Distillate and High Density Fuels. Report ESL-TR-81-54. Tyndall Air Force Base, Fla.: Engineering and Services Laboratory, Air Force Engineering and Services Center.

Wilson, J. T., J. M. Armstrong, and H. Rifai. In press. A full scale field demonstration on the use of hydrogen peroxide for in-situ bioremediation of an aviation gasoline-contaminated aquifer. In Bioremediation: Field Experiences, P. E. Flathman, D. E. Jerger, and J. H. Exner, eds. Chelsea, Mich.: Lewis Publishers.

Zapico, M. M., S. Vales, and J. A. Cherry. 1987. A wireline piston core barrel for sampling cohesionless sand and gravel below the water table. Ground Water Monitoring Review 7(3):74-82.

Appendixes

A

Glossary

Abiotic—Occurring without the involvement of microorganisms.

Aerobic respiration—Process whereby microorganisms use oxygen as an electron acceptor to generate energy.

Air sparging—Injection of air into ground water to remove volatile chemicals and deliver oxygen, which promotes microbial growth.

Air stripping—Above-ground process used to remove volatile contaminants from water. It involves exposing the water surface to a large volume of air, usually by flowing water through a tower in one direction and air through the tower in the opposite direction.

Aliphatic hydrocarbon—A compound built from carbon and hydrogen joined in a linear chain. Petroleum products are composed primarily of aliphatic hydrocarbons.

Anaerobic respiration—Process whereby microorganisms use a chemical other than oxygen as an electron acceptor. Common "substitutes" for oxygen are nitrate, sulfate, and iron.

Aquifer—An underground geological formation that stores ground water.

Aromatic hydrocarbon—A chemical formed from benzene rings, originally called "aromatic" because of benzene's distinctive aroma. Solvents, many types of pesticides, and polychlorinated biphenyls are composed of aromatic hydrocarbons.

Bacterium—A single-celled organism of microscopic size. Bacteria are ubiquitous in the environment, inhabiting water, soil, organic matter, and the bodies of plants and animals.

Benzene—A chemical composed of six carbon atoms arranged in a hexagonal ring, with one hydrogen atom attached to each carbon.

Bioaugmentation—The addition of nonnative microorganisms to a site.

Biocurtain—A large quantity of organisms grown underground specifically to stop contaminant migration by creating localized clogging.

Biodegradation—Biologically mediated conversion of one compound to another.

Biomass—Total mass of microorganisms present in a given amount of water or soil.

Bioremediation—Use of microorganisms to control and destroy contaminants.

Biotransformation—Microbially catalyzed transformation of a chemical to some other product.

Bioventing—Circulation of air through the subsurface to remove volatile contaminants and provide oxygen, which stimulates microorganisms to degrade remaining contaminants.

BTEX—Acronym for benzene, toluene, ethylbenzene, and xylenes, which are compounds present in gasoline and other petroleum products, coal tar, and various organic chemical product formulations.

Carbon treatment—Above-ground process for removing contaminants from water or air. It involves contact between the water or air and activated carbon, which adsorbs the contaminants, usually by flowing the water or air through columns packed with carbon.

Carbonate—Any chemical containing the CO_3^{2-} group; limestone and dolomite are examples of rocks formed primarily from carbonate minerals.

Chlorinated solvent—A hydrocarbon in which chlorine atoms substitute for one or more hydrogen atoms in the compound's structure. Chlorinated solvents commonly are used for grease removal in manufacturing, dry cleaning, and other operations. Examples include trichloroethylene, tetrachloroethylene, and trichloroethane.

Cometabolism—A reaction in which microbes transform a contaminant even though the contaminant cannot serve as an energy source for the organisms. To degrade the contaminant, the microbes require the presence of other compounds (primary substrates) that can support their growth.

Complexing agent—A chemical agent that chemically bonds with a positively charged molecule, such as a metal. Complexing agents can be used to dissolve precipitated metals.

Conservative tracer—A chemical that does not undergo microbiological reactions but has transport properties similar to those of microbiologically reactive chemicals (such as the contaminant and oxygen).

Dechlorinate—The removal of chlorine atoms from a compound.

Desorption—Opposite of sorption; the dissolution of chemicals from solid surfaces.

Deuterium—Hydrogen isotope with twice the mass of ordinary hydrogen; it contains one proton and one neutron in its nucleus.

Diauxy—Selective biodegradation of some organic compounds over others, which sometimes occurs when the compounds are present in mixtures.

DNA (deoxyribonucleic acid)—Substance within a cell that passes hereditary information from one generation to the next.

Electron—A negatively charged subatomic particle that may be transferred between chemical species in chemical reactions. Every chemical molecule contains electrons and protons (positively charged particles).

Electron acceptor—Compound that receives electrons (and therefore is reduced) in the energy-producing oxidation-reduction reactions that are essential for the growth of microorganisms and bioremediation. Common electron acceptors in bioremediation are oxygen, nitrate, sulfate, and iron.

Electron donor—Compound that donates electrons (and therefore is oxidized) in the energy-producing oxidation-reduction reactions that are essential for the growth of microorganisms and bioremediation. In bioremediation the organic contaminant often serves as an electron donor.

Engineered bioremediation—Type of bioremediation that stimulates the growth and biodegradative activity of microorganisms by adding nutrients, electron acceptors, or other stimulants to the site using an engineered system.

Enzyme—A protein created by living organisms to use in transforming a specific compound. The protein serves as a catalyst in the compound's biochemical transformation.

Enzyme induction—Process whereby an organism synthesizes an enzyme in response to exposure to a specific chemical, the inducer.

Equilibrium—Condition in which a reaction has occurred to its maximum extent.

Ex situ—Latin term referring to the removal of a substance from its natural or original position.

Fermentation—Process whereby microorganisms use an organic compound as both electron donor and electron acceptor, converting the compound to fermentation products such as organic acids, alcohols, hydrogen, and carbon dioxide.

Fixation—Process whereby microorganisms obtain carbon for building new cells from inorganic carbon, usually carbon dioxide.

Free product recovery—Removal of residual pools of contaminants, such as gasoline floating on the water table, from the subsurface.

Gas chromatograph—Instrument used to identify and quantify volatile chemicals in a sample.

Gene probe—One class of oligonucleotide probes. Gene probes are used to identify the presence of a particular gene (such as the gene responsible for a particular biodegradative reaction) on the cell's DNA.

Genetically engineered organism—An organism whose genes have been altered by humans. For example, researchers have used genetic engineering to give bacteria the capability to degrade hazardous chemicals that normally resist biodegradation.

Glacial outwash—Materials (typically sand and gravel) deposited during the melting of glaciers.

Halogenate—Replacement of one or more hydrogen atoms on a chemical compound with atoms of a halogen, such as chlorine, fluorine, or bromine.

High-performance liquid chromatograph—Instrument used to identify and quantify contaminants in a sample.

Hydraulic conductivity—A measure of the rate at which water moves through a unit area of the subsurface under a unit hydraulic gradient.

Hydraulic gradient—Change in head (i.e., water pressure) per unit distance in a given direction, typically in the principal flow direction.

Hydrocarbon—A chemical composed of carbon and hydrogen in any of a wide variety of configurations. Petroleum products, as well as many synthetic industrial chemicals, contain many different hydrocarbons.

Hydrophobic compound—A "water-fearing" compound, such as oil, that has low solubility in water and tends to form a separate phase.

In situ—Latin term meaning "in place"—in the natural or original position.

Infiltration gallery—Engineered system used to deliver materials that stimulate microorganisms in the subsurface. Infiltration galleries typically consist of buried perforated pipes through which water containing the appropriate stimulating materials is pumped.

Inorganic compound—A chemical that is not based on covalent carbon bonds. Important examples are metals, nutrients such as nitrogen and phosphorus, minerals, and carbon dioxide.

Intrinsic bioremediation—A type of in situ bioremediation that uses the innate capabilities of naturally occurring microbes to degrade contaminants without taking any engineering steps to enhance the process.

Intrinsic permeability—A measure of the relative ease with which a liquid will pass through a porous medium. Intrinsic permeability depends on the shape and size of the openings through which the liquid moves.

Isotope—Any of two or more species of an element in the periodic table with the same number of protons. Isotopes have nearly identical chemical properties but different atomic masses and physical properties. For example, the isotope carbon 12 has six protons and six neutrons, while the isotope carbon 13 has six protons and seven neutrons. Both have atomic number 6 (the number of protons), but carbon 13 is more massive than carbon 12 because it carries an extra neutron.

Isotope fractionation—Selective degradation by microorganisms of one isotopic form of a carbon compound over another isotopic form. For example, microorganisms degrade the ^{12}C isotopes of petroleum hydrocarbons more rapidly than the ^{13}C isotopes.

Kinetics—Refers to the rate at which a reaction occurs.

Land farming—Above-ground process used to stimulate microorganisms to degrade contaminants in soil. The process involves spreading out the soil, adding nutrients, and tilling.

Ligand—See "complexing agent."

Mass spectrometer—Instrument used to identify the chemical structure of a compound. Usually, the chemicals in the compound are separated beforehand by chromatography.

Metabolic intermediate—A chemical produced by one step in a multistep biotransformation.

Metabolism—The chemical reactions in living cells that convert food sources to energy and new cell mass.

Methanogen—A microorganism that produces methane. Because they thrive without oxygen, methanogens can be important players in subsurface biotransformations, where oxygen is often absent.

Micelle—An aggregate of molecules, such as surfactant molecules, that form a small region of nonaqueous phase within an otherwise aqueous matrix.

Microcosm—A laboratory vessel set up to resemble as closely as possible the conditions of a natural environment.

Microorganism—An organism of microscopic or submicroscopic size. Microorganisms can destroy contaminants by using them as "food sources" for their own growth and reproduction.

Mineralization—The complete degradation of an organic chemical to carbon dioxide, water, and possibly other inorganic compounds.

Most-probable-number (MPN) technique—A statistical technique for estimating the number of organisms present in a sample.

Nonaqueous-phase liquid—A liquid solution that does not mix easily with water. Many common ground water contaminants, including chlorinated solvents and many petroleum products, enter the subsurface in nonaqueous-phase solutions.

Oligonucleotide probe—A short piece of DNA that can be used to identify the genetic makeup of microorganisms in a sample and the reactions they are capable of carrying out.

Organic compound—A compound built from carbon atoms, typically linked in chains or rings.

Oxidization—Transfer of electrons away from a compound, such as an organic contaminant. The oxidation can supply energy that microorganisms use for growth and reproduction. Often (but not always), oxidation results in the addition of an oxygen atom and/or the loss of a hydrogen atom.

Petroleum hydrocarbon—A chemical derived from petroleum by various refining processes. Examples include gasoline, fuel oil, and a wide range of chemicals used in manufacturing and industry.

Plume—A zone of dissolved contaminants. A plume usually originates from the contaminant zone and extends for some distance in the direction of ground water flow.

Primary substrates—The electron donor and electron acceptor that are essential to ensure the growth of microorganisms. These compounds can be viewed as analogous to the food and oxygen that are required for human growth and reproduction.

Protozoan—A single-celled organism that is larger than a bacterium and may feed on bacteria.

Pump-and-treat system—Most commonly used type of system for cleaning up contaminated ground water. Pump-and-treat systems consist of a series of wells used to pump contaminated water to the surface and a surface treatment facility used to clean the extracted ground water.

Rate-limiting material—Material whose concentration limits the rate at which a particular process can occur.

Reduction—Transfer of electrons to a compound, such as oxygen. It occurs when another compound is oxidized.

Reductive dehalogenation—A variation on biodegradation in which microbially catalyzed reactions cause the replacement of a halogen atom on an organic compound with a hydrogen atom. The reactions result in the net addition of two electrons to the organic compound.

Reporter gene—A tool used with genetically engineered microorganisms. When a reporter gene is incorporated into a microorganism's genetic material, it provides a signal when the organism is present and active. An example is a gene that produces a protein that causes the microorganism to emit light.

Saturated zone—Part of the subsurface that is beneath the water table and in which the pores are filled with water.

Secondary substrate—A chemical that can be transformed by microorganisms through secondary utilization.

Secondary utilization—General term for the transformation of contaminants by microorganisms when the transformation yields little or no benefit to the organisms.

Slurry wall—A clay barrier constructed in the subsurface to prevent the spread of contaminants by preventing water flow.

Soil vapor extraction—See "Vapor recovery."

Sorption—Collection of a substance on the surface of a solid by physical or chemical attraction.

Substrate—A compound that microorganisms can use in the chemical reactions catalyzed by their enzymes.

Sulfate reducer—A bacterium that converts sulfate to hydrogen sulfide. Because they can act without oxygen, sulfate-reducing bacteria can be important players in the oxygen-limited subsurface.

Surfactant—Soap or a similar substance that has a hydrophobic and a hydrophilic end. Surfactants can bond to oil and other immiscible compounds to aid their transport in water.

Unavailability—Situation in which a contaminant is sequestered from the microorganism, inhibiting the organism's ability to degrade the contaminant.

Unsaturated zone—Soil above the water table, where pores are partially or largely filled with air.

Vadose zone—See "Unsaturated zone."

Vapor recovery—A method for removing volatile contaminants from the soil above the water table by circulating air through the soil.

Volatilization—Transfer of a chemical from the liquid to the gas phase (as in evaporation).

B

Biographical Sketches of Committee Members and Staff

BRUCE E. RITTMANN, committee chair, is the John Evans Professor of Environmental Engineering at Northwestern University and an active researcher and teacher in the field of environmental biotechnology. His special interests include biofilm kinetics, microbial ecology, in situ bioremediation, biological drinking water treatment, and the fate of hazardous organic chemicals. Dr. Rittmann was a member of the National Research Council committee that authored *Ground Water Models: Scientific and Regulatory Applications*. He recently served as president of the Association of Environmental Engineering Professors.

LISA ALVAREZ-COHEN, assistant professor of environmental engineering at the University of California, Berkeley, received a Ph.D. in environmental engineering and science from Stanford University in 1991 and a B.A. in engineering and applied science from Harvard University in 1984. Dr. Alvarez-Cohen's research interests are modeling of microbial processes in porous media, bioremediation of contaminated aquifers, innovative hazardous waste treatment technologies, and application of cometabolic biotransformation reactions.

PHILIP B. BEDIENT, Shell Distinguished Professor of Environmental Science at Rice University, received a B.S. in physics in 1969, an M.S. in environmental engineering in 1972, and a Ph.D. in environmental engineering sciences in 1975 from the University of Florida.

His primary research interests include ground water pollutant transport modeling and hazardous waste site evaluation.

RICHARD A. BROWN, vice president of remediation technology for Groundwater Technology, Inc., in Trenton, New Jersey, received a B.A. in chemistry from Harvard University and a Ph.D. in inorganic chemistry from Cornell University. His responsibilities include the development and implementation of remediation technologies such as bioremediation, soil vapor extraction, and air sparging. Before joining Groundwater Technology, Dr. Brown was director of business development for Cambridge Analytical Associates' Bioremediation Systems Division and technology manager for FMC Corporation's Aquifer Remediation Systems. Dr. Brown holds patents on applications of bioreclamation technology, on the use of hydrogen peroxide in bioreclamation, and on an improved nutrient formulation for the biological treatment of hazardous wastes.

FRANCIS H. CHAPELLE, a researcher at the U.S. Geological Survey in Columbia, South Carolina, received a Ph.D. in hydrology in 1984 from George Washington University. He also holds a B.A. in music and a B.S. in geology from the University of Maryland. Currently, he studies the impacts of subsurface microbiology on ground water chemistry.

PETER K. KITANIDIS, professor of civil engineering at Stanford University, received a B.S. in civil engineering in 1974 from the National Technical University of Athens, Greece, an M.S. in civil engineering in 1976 from the Massachusetts Institute of Technology, and a Ph.D. in water resources in 1978, also from MIT. His current research focuses on the use of geostatistical and predictive ground water hydrology methods for designing water quality monitoring networks. He is also conducting research on the design of nutrient circulation systems for stimulating subsurface microorganisms to degrade ground water contaminants.

EUGENE L. MADSEN, assistant professor in the Section of Microbiology, Division of Biological Sciences, at Cornell University, received B.A., B.S., and M.S. degrees from the University of California at Santa Cruz, Oregon State University, and Cornell University, respectively. His Ph.D. from Cornell in 1985 is in soil science, microbiology, and ecology. Since 1989, as a researcher at Cornell, he has pursued interests in ground water microbiology, microbial metabolism of environmental pollutants, and developing criteria for proving in situ biodegradation. Prior to returning to Cornell, he held research appointments at Rutgers and Penn State universities and at an environmental restoration company in Bozeman, Montana.

WILLIAM R. MAHAFFEY, vice president of the technology department for ECOVA Corporation in Redmond, Washington, earned a B.S. in microbiology in 1976 and an M.S. in microbial ecology in 1978 from the State University of New York. He holds a Ph.D. in microbial biochemistry, earned in 1986, from the University of Texas. Dr. Mahaffey serves as technical principal on all of ECOVA's bioremediation projects. He directs the activities of project microbiologists, develops operating parameters for the use of enhanced biodegradation in the field, and reviews all company projects involving bioremediation.

ROBERT D. NORRIS, technical director of bioremediation at Eckenfelder, Inc., in Nashville, Tennessee, received a B.S. in chemistry from Beloit College and a Ph.D. in organic chemistry from the University of Notre Dame. He has been involved since 1983 in the development and implementation of a variety of bioremediation processes and holds 13 U.S. patents, including four on various aspects of bioremediation. He has developed and conducted laboratory and pilot tests as well as successful in situ and ex situ bioremediation under a range of conditions.

JOSEPH P. SALANITRO, senior staff research microbiologist at Shell Development Company in Houston, Texas, holds a Ph.D. in microbiology from Indiana University. During his 21-year career with Shell, he has been involved in both the chemical and oil sectors of environmental research, studying the aerobic and anaerobic biodegradability of detergents, chemicals, pesticides and petrochemical waste effluents, and the role of microbes and sour gas formation in oil field waterfloods. His current research interests are in defining the potential and limits of biodegradation of gasoline components in subsurface remediation.

JOHN M. SHAUVER, environmental enforcement manager for the Michigan Department of Natural Resources, earned a B.S. in fisheries biology from Michigan State University and did two years of postgraduate work in geology. A 24-year veteran of the Department of Natural Resources (minus a two-year absence for military duty from 1968 until 1970), he has worked as a water quality investigator, hazardous waste cleanup specialist, aquatic biologist, environmental law enforcement specialist, and environmental enforcement manager.

JAMES M. TIEDJE, director of the Center for Microbial Ecology at Michigan State University, received a B.S. in agronomy in 1964 from Iowa State University. He received an M.S. in 1966 and a Ph.D. in 1968 in soil microbiology from Cornell University. He is currently a professor in the Departments of Crop and Soil Sciences and Microbiology and Public Health at Michigan State University. His expertise is in microbial ecology, and he conducts research in three focal areas: denitrification, reductive dehalogenation, and the use of gene probes to study community selection in nature.

JOHN T. WILSON, research microbiologist at the U.S. Environmental Protection Agency's R. S. Kerr Laboratory in Ada, Oklahoma, earned a B.S. in biology in 1969 from Baylor University, an M.A. in microbiology in 1971 from the University of California at Berkeley, and a Ph.D. in microbiology in 1978 from Cornell University. His areas of expertise are bioremediation and subsurface microbiology, with emphasis on quantitative description of the biological and physical processes that control the behavior of hazardous materials in soils and the subsurface.

RALPH S. WOLFE, a professor of microbiology at the University of Illinois since 1955, received a Ph.D. at the University of Pennsylvania in 1953. A member of the National Academy of Sciences, his major research interest has been anaerobic microbial metabolism. He was attracted to the study of methanogenic bacteria in 1961 because their biochemistry was unknown, and the difficulty of isolating and cultivating these extremely oxygen-sensitive anaerobes was legendary. Dr. Wolfe developed a system for mass culture of methanogens in kilogram quantities, obtained the first formation of methane by a cell-free extract, and evolved a simplified procedure for the routine culture of methanogens in a pressurized atmosphere of hydrogen and carbon dioxide—a technique that has played a pivotal international role in development of the field.

JACQUELINE A. MACDONALD, program officer at the National Research Council's Water Science and Technology Board, served as study director and managing editor for the Committee on In Situ Bioremediation. She holds an M.S. in environmental science in civil engineering from the University of Illinois and a B.A., *magna cum laude*, in mathematics from Bryn Mawr College.

GREGORY K. NYCE, senior project assistant at the National Research Council's Water Science and Technology Board, served as project assistant for the Committee on In Situ Bioremediation. He received his B.S. in psychology from Eastern Mennonite College.

Index

I

J

K

L

M

T

U

V

W

X

Other Recent Reports of the Water Science and Technology Board

Ground Water Vulnerability Assessment: Predicting Contamination Potential Under Conditions of Uncertainty (1993)
Managing Wastewater in Coastal Urban Areas (1993)
Sustaining Our Water Resources: Proceedings, WSTB Symposium (1993)
Water Transfers in the West: Efficiency, Equity, and the Environment (1992)
Restoration of Aquatic Ecosystems: Science, Technology, and Public Policy (1992)
Toward Sustainability: Soil and Water Research Priorities for Developing Countries (1991)
Preparing for the Twenty-first Century: A Report to the USGS Water Resources Division (1991)
Opportunities in the Hydrologic Sciences (1991)
A Review of the USGS National Water Quality Assessment Pilot Program (1990)
Ground Water and Soil Contamination Remediation: Toward Compatible Science, Policy, and Public Perception (1990)
Managing Coastal Erosion (1990)
Ground Water Models: Scientific and Regulatory Applications (1990)
Irrigation-Induced Water Quality Problems: What Can Be Learned from the San Joaquin Valley Experience? (1989)

Copies of these reports may be ordered from
the National Academy Press
1-800-624-6242
202-334-3313

In situ bioremediation